Systems-of-Systems Perspectives and Applications - Design, Modeling, Simulation and Analysis (MS&A), Gaming and Decision Support

Edited by Tien M. Nguyen

Published in London, United Kingdom

IntechOpen

Supporting open minds since 2005

Systems-of-Systems Perspectives and Applications - Design, Modeling, Simulation and Analysis (MS&A), Gaming and Decision Support
http://dx.doi.org/10.5772/intechopen.77586
Edited by Tien M. Nguyen

Contributors
Andrzej Łodziński, Sameh Najeh, Ammar Bouallegue, Teresa Azevedo Perdicoúlis, Paulo Lopes Dos Santos, Tien M. Manh Nguyen, Alper Pahsa, Thomas O. Freeze, John D. Nguyen, Maria Cerreta, Ludovica La Rocca, Gaia Daldanise, Paolo Di Lauro, Fatima Isiaka, Awwal M. Adamu, Salihu A. Abdulkarim, Ruolun Liu, Xueqin Zhang, Rui Huang, Lucia Pusillo, Agostino Bruzzone, Derseh Limeneh, Kelem Yilma, Tung Bui, Charles H. Lee, Abdullahi Salihu, G. Jank

First published in London, United Kingdom, 2021 by IntechOpen
IntechOpen is the global imprint of INTECHOPEN LIMITED, registered in England and Wales, registration number: 11086078, 5 Princes Gate Court, London, SW7 2QJ, United Kingdom
Printed in Croatia

British Library Cataloguing-in-Publication Data
A catalogue record for this book is available from the British Library

Additional hard and PDF copies can be obtained from orders@intechopen.com

Systems-of-Systems Perspectives and Applications - Design, Modeling, Simulation and Analysis (MS&A), Gaming and Decision Support
Edited by Tien M. Nguyen
p. cm.
Print ISBN 978-1-83968-327-5
Online ISBN 978-1-83968-328-2
eBook (PDF) ISBN 978-1-83968-329-9

Meet the editor

Dr. Tien Nguyen serves as an adjunct research professor in Mathematics at CSUF, where he is also a visiting scholar and advisor at the Center of Computational and Applied Mathematics. He works full-time as a deputy chief architect of an International Program at The Aerospace Corporation. Prior to this, he was associate director, interim director, and principal technical staff. He was with Raytheon serving as Program-Area Chief Engineer in Advanced Concept Technology and retired as Engineering Fellow in 2014. He also served as NASA delegate to the international CCSDS and many of his works were adopted as CCSDS standards. He received his Ph.D. in Applied Mathematics from the Claremont Graduate University. He is an expert in advanced mathematical modeling of complex systems-of-systems.

Contents

Preface

This professional technical book is intended for engineers, scientists, and researchers in both systems engineering advanced mathematical modeling and simulation fields. The objectives of this book are to present a collection of chapters on (i) Systems-of-Systems (SOS) engineering and SOS architecture design and analysis addressing the challenges associated with SOS Enterprise Architecture (SOSEA) design and SOS Enterprise (SOSE) CONOPS Modeling, Simulation and Analysis (MS&A); (ii) SOS perspectives with MS&A applications on practical systems; and (iii) decision support systems and game theoretical modeling and analysis. To meet these objectives, this book is organized as follows:

- Section 1 "Recent Trends in SOS Design and MS&A": This section provides an introductory chapter discussing recent trends in SOS architecture design, MS&A for complex systems, and gaming and decision support systems, and it relates these trends to the chapters presented in this book. This introductory chapter also discusses the differences between Systems-of-Systems (SOS) and System-of-Systems (SoS) and the extension of the existing SoS engineering approach to SOS for architecture design and MS&A of complex systems or processes.

- Section 2 "Systems-of-Systems (SOS) Engineering": This section includes two chapters, namely, Chapters 2 and 3, with emphasis on the taxonomy of a complex SOS and System-of-Systems (SoS) engineering method and process for managing an enterprise transformation. Chapter 2 provides (i) a thorough discussion of SOS taxonomy with emphasis on space and airborne systems applications and (ii) two examples of military satellite communications (SATCOM) and manned aircraft systems. Chapter 3 discusses the applications of system engineering technical process flows on enterprise systems with an emphasis on SoS perspectives.

- Section 3 "SOS Architecture Design and Analysis": This section also provides two chapters. Chapter 4 defines the SOSE, SOSEA, and SOSE CONOPS and presents an approach for the design and development of a complex SOSEA with a focus on space and airborne systems. Chapter 4 also presents SOSEA CONOPS assessment framework and models. Chapter 5 discusses the implementation of the proposed SOS framework and MS&A models in MATLAB and presents preliminary results for a notional space SOSEA in the presence of radio frequency interference (RFI). Chapter 5 also presents newly proposed SOSEA resiliency metrics for assessing complex SOSE CONOPS, including Resilient Assessment Index Against RFI (RAI-RFI), Spectrum Resiliency Assessment Index (SRAI), and Resilient Capacity (RC).

- Section 4 "SOS Modeling, Simulation and Analysis (MS&A) Applications": This section presents three chapters, namely, Chapters 6, 7, and 8, demonstrating the use of SoS and SOS perspectives to simulate and analyze practical complex processes and systems related to the melt spinning process, and test systems for assessing the effect of friction on drive screw and spectrum analysis systems for estimating frequencies of multiple linear frequency modulated (LFM) signals. Chapter 6 discusses the use of SoS engineering for the development of computer models to simulate a complex melt spinning process (MSP) and presents a new SoS perspective on MS&A of MSP to reduce the risk for achieving uniformity of the fiber quality. Chapter 7 presents a new approach in SOS engineering to (i) develop computer models for simulating the effect of friction on a drive screw with a predetermined torque and (ii) develop MS&A models of the drive screw. Chapter 7 also uses the SOS perspective from a control engineering viewpoint to examine the complex drive screw control process (DSCP) to connect the SOS MS&A models and simulate the DSCP process flow for achieving optimum performance. Chapter 8 demonstrates the use of SOS perspective and standard SoS engineering approach for the decomposition of the frequency estimation process (FEP) into three systems, including System 1 responsible for transforming the time-domain signal into a frequency-domain signal, and time-frequency (TF) transform using continuous wavelet transform (CWT) with a focus on the chirplet transform (CT); System 2 responsible for detecting instantaneous signal frequency (IF) in the presence of noise; and (c) System 3 responsible for assessing of the filter impulse response length on TF distribution and analyzing the TF behavior and applying Hough transform (HT) for frequency estimation. Furthermore, Chapter 8 also presents simulation results for estimating frequencies of multiple LFM signals.

- Section 5 "Decision Support Systems (DSS) Applications": There are two chapters in this section, namely, Chapters 9 and 10. Chapter 9 presents an approach for decision support under risk using multiple criteria optimizations. This chapter discusses the difficulty in assigning values to decision parameters and proposes an iterative process in discovering preferences and fine-tuning personal aspiration levels. It shows the benefits of this iterative process to help the decision-makers assess their objectives during the decision-making process and improve the consistency between the iterative evolution of the analysis and the changes in the decision-maker's preferences. Chapter 10 reports the on-going experiment with Play ReCH (Reuse Cultural Heritage) as a case study of how collaborative decision-making can help promote cooperation and innovation in the context of urban public policy. The chapter reports that early lessons learned from the Play ReCH project have shown a strong potential for a reduction of cultural poverty, a greater social cohesion, and perpetuation of the culture to the next generation.

- Section 6 "Game Theory Applications": This section addresses a wide range of game theoretical models (GTM), including theoretical disturbed Stackelberg games, innovative distributed control approach using game theory tools for mode selection and power control in 5G wireless communications device-to-device (D2D), and game theory applications for modeling, simulation and analysis of the evolution of tumor cells, neocortical epilepsy, schizophrenia, brain disorders, and cancer. To address these game theory applications, there are three chapters in this section: Chapters 11, 12, and 13. Chapter 11 offers a generalization of the Stackelberg game that helps antagonists reach an equilibrium through an open loop process. Chapter 11 confirms the results of some earlier studies that, using symmetric Riccati differential equations, a convexity would warrant the existence of a unique Stackelberg equilibrium. With the ever-growing wireless traffic in D2D communication, users are competing with an increasingly unavailable spectrum causing interference.

Chapter 12 presents a noncooperative game for managing power control in D2D uplink underlaid cellular networks. This chapter discusses a noncooperative game model setup and shows the model iteratively negotiates a min-max range of power that leads to Nash equilibrium. Finally, Chapter 13 addresses game theory that can be appropriately applied to healthcare. This chapter discusses the application of game theory to forecast the evolution of tumor cells, neocortical epilepsy surgery, schizophrenia, and cancer.

The remaining task of this preface is to recognize the contributions of key people for the creation of this professional technical book. I would like to express my deep gratitude to all the authors and co-authors for their contributions. The success of this book has not only been the result of the work from the authors and coauthors but also from the cooperation of many people, including numerous reviewers and the people at our IntechOpen publisher who provided constant support. Particularly, I would like to thank (i) the IntechOpen Author Service Manager, Ms. Romina Rovan, for her invaluable assistance, conscientious, and relentless support during the editing and publishing process—without her this book is not possible; (ii) former IntechOpen Commissioning Editor, Mrs. Klara Mestrovic, and the new Commissioning Editor, Ms. Jenela Germuth, for their continuous support; (iii) Professor Carlos Pedro Gonçalves of Universidade Lusófona de Humanidades e Tecnologias for his review and editing of several chapters on game theory applications; (iv) Professor Tung X. Bui of University of Hawaii for his review of decision support and game theory chapters and valuable contributions to introductory Chapter 1, and Professor Hien Tran of North Carolina State University for his review of SOS chapters; (v) my colleagues at California State University in Fullerton, Professors Charles Lee and Sam Behseta for their continuous support; and (vi) my colleagues and managers at The Aerospace Corporation, Dr. Hung Nguyen, Mr. Andy Guillen, Dr. Sumner Matsunaga, Navneet Mezcciani, and Garick Leu-cheung for their constant support and review of SOS chapters. Finally, I'm forever indebted to my wonderful wife, Thu-Hang Nguyen, for her moral support and constant encouragement during the process of making this complex and diverse professional technical book.

Tien M. Nguyen
Adjunct Research Professor,
Visiting Scholar, Center for Computational and Applied Mathematics (CCAM),
California State University in Fullerton (CSUF),
Fullerton, California, U.S.A
(also, with The Aerospace Corporation)

Section 1

Recent Trends in SOS Design and MS&A

Chapter 1

Introductory Chapter: Recent Trends in Systems-of-Systems Design, Modeling, Simulation and Analysis for Complex Systems, Gaming and Decision Support Final

Tien M. Nguyen and Tung X. Bui

1. Introduction

This book is a collection of some of the latest development work that embrace three major interrelated scientific disciplines – systems theory, decision theory and game theory, and the information and communication science with the goal to deploy innovative and effective computer-support application Systems of Systems (SOS). In addition, it also suggests the adoption of the Decision Support as a process-driven approach to iteratively engineer SOS. The book is a collective effort that uses a diverse set of studies using a wide spectrum of SOS applications to shed insights in the use of MS&A models to assist decision makers. Thanks to the exponential growth of big data related to all aspects of human activities, the surge in decision-making complexity due to the current climate of uncertainty with unforeseen consequences, and the increasing pervasiveness of advanced information and communication technologies (ICT) such as the proliferation of mobile apps, Internet-of-Things (IoT) and bots, we have witnessed an acceleration of integration of "complex systems" across of a wide range of application domains that include, but are not limited to, telecommunications, medicine (healthcare), military, manufacturing, transportation, energy, social networking platforms, education, and arts and culture. From Reference [1] the perspectives of system engineering, complex systems can be classified as:

- SOS Type 1: A family of system-of-systems that provides similar core services, e.g., communication services. But each system provides different core service types, e.g., non-secure FDMA vs. secure TDMA[1] communication services;

- SOS Type 2: An integration of many families of SoS. When combined, this type of system provides unique SOS capabilities at the enterprise level (i.e., integrated level). An example of this complex system is a combination of a family of communications SoS with a family of Global Position Satellite (GPS) SoS; and

[1] FDMA = Frequency Division Multiple Access vs. TDMA = Time Division Multiple Access.

- SOS Type 3: An integration of many heterogenous, independent but interrelated types of systems with each system providing distinctive core services. For example, a production line consists of (i) electrical system, (ii) sensor system, and (iii) mechanical system with belt conveyor, etc.

Note that we use U.S. Department of Defense (DoD) SOS System engineering guide [1] to define SOS. While many existing papers, documents and System-of-Systems (SoS, which is not SOS) standards considered integration of (i) many systems of the same type of systems together which is identical to our Type 1, and (ii) many different types of systems as a system consisted of many systems and referred to as SoS, which is identical to our SOS Type 3. In this chapter, we focus our discussion on SOS Type 2, since existing SoS engineering standards can be directly applied to SOS Type 1 and Type 3 but not Type 2.

It would be safe to claim that the disciplines of current system thinking, decision science and computer and telecommunication engineering have played a critical role in the emergence of SoS[2] and SOS. Traditional system theory posits that the whole is greater than the sum of its isolated parts [2–4]. And to achieve this holistic added value, it seeks to identify, analyze and create processes through optimal arrangements or adaptations of individual and independent subsystems' components and systems under system and SoS perspectives, respectively. This part-to-whole and whole-to-part thinking has been extended to SOS perspective for complex systems belong to SOS Type 2 and is prevalent in the selected chapters of this book.

Decision theory deals with the reasoning -- be it rational or not -- that drives a person's choice. The three core concepts in decision theory are elicitation and interpretation of the decision maker's preferences, the search of available options, and the management of uncertainty, risks and regrets [5–7]. In organizational or collective settings, decision making is extended to multiple stakeholders. Von Neumann, Morgenstern and Nash are universally credited for their pioneering work on game theory [8–10]. They proposed mathematical models of strategic interaction among rational decision-makers. The latter can be either cooperative or non-cooperative. The discussion on modern SOS in this volume constantly the basic foundations of game theory that uses Nash equilibrium as a prime operational goal.

The third underlying discipline that unifies the chapter of this book is the discipline of computer, information, communication and computer system (ICS) engineering. As ICS engineers continue to stay at the forefront of technological development, specific issues related to process flow design and conflict management of federated systems have emerged to be most challenging. The majority of chapters in this book, throughout their specific domain applications, such as space systems, attempt to address the challenges related to merging computing and communications due to the limitations of existing core protocols and at SOS design level, a deep understanding of how to conceptualize, design and manage coordinating parallelism [11, 12]. These design issues appear virtually in all chapters of this book. And the authors have demonstrated several different approaches to address these challenges in their specific applications.

As mentioned earlier, the authors use the DoD Systems Engineering Guide published in 2008 for defining SOS [1]. U.S. DoD has defined SOS as "a set or arrangement of systems that results when independent and useful systems are integrated into a larger system that delivers unique capabilities." SOS development involves the creation of systems, which are collections of legacy, evolving and new systems that must have a high degree of flexibility and adaptability. Two of the main considerations for defining SOS are:

[2] Note that in this Chapter SoS can be considered as SOS Type 1 and Type 2 but not Type 3. Existing standard SoS engineering approach needs to be modified or extended to address SOS Type 2 enterprise challenges.

i. The lack of authority over the constituent systems because of their independent management, funding and objectives that may not align with those of the SOS as a whole.

ii. The emergent behavior adds a large degree of unpredictability, as overall system behavior cannot be predicted by having individual knowledge of each of the constituents.

Recently in 2019, the International Council on Systems Engineering (INCOSE) [13] has released the latest version of the "Guidelines for the Utilization of ISO/IEC/IEEE 15288 in the Context of System of Systems (SoS) Engineering", which is an extension of existing system engineering, to industry for review and comments. This document provides guidance for the utilization of ISO/IEC/IEEE 15288 in the context of SoS in many domains. This INCOSE guidebook perceives that SoS engineering demands a balance between linear procedural procedures for systematic activity and holistic nonlinear procedures due to additional complexity from SoS emergence. ISO/IEC/IEEE 15288 is considered by most system engineers as the foundation for SoS Engineering for many civilian and commercial applications, while U.S. DoD Systems Engineering Guide for SOS is considered as the foundation for SOS engineering design and implementation for defense applications.

The following sections address the objectives presented in the abstract section.

2. Current trends on SOS engineering

Using a requirement-based approach, existing System-of-Systems (SoS) engineering is a software engineering discipline that deals with the analysis, design, development, deployment and evaluation of heterogenous systems. A peculiar aspect of SoS engineering is to deal with a significant level of uncertainty in requirements engineering [13]. As such, while the goal of traditional system engineering is to build the system right, SoS engineering goal is to build the right system in the SoS context. Instead of optimizing individual systems, the primary objective of SoS engineering methodology is to maximize the overall performance of an integrated platform.

Many researchers, scientists and system engineers in the field have recognized the following challenges related to existing requirement based SoS approach when extended to complex SOS Type 2 defined above: (i) Requirements are usually derived from the assumptions of certain selected technology enablers that are usually a few years behind the current technologies, and (ii) When the requirements are changing at a fast pace, the current (As-Is) system architecture design using this SoS requirement-based approach becomes a "bottleneck" when interfacing with newer (To-Be) systems that are not within the same family. The current SOS engineering approach is to focus on the use of capability-based SOS engineering approach to address the "requirements" challenges for monolithic and complex systems (SOS Type 2 or complex SOS) with changing requirements due to technology changes and dynamic environments.

3. Current trends on SOS architecture design and MS&A

Known as complex SOS, these new deployments represent a step further in integrating task-driven dedicated systems that pool and share resources (i.e., data) and capabilities (i.e., modeling and intelligence) together to expand more functionalities and improved SOS performance (i.e., effectiveness and efficiency) to deliver unique capabilities than simply the sum of the individual constituent systems. Many of these SOS appear invisible to most users, but their presence is omnipresent in our daily lives: smart power grids, integrated traffic networks for air, land and sea connectivity, health information systems, global supply chain networks, and many others.

The two key parts of using existing standard requirement based SoS engineering for extension to SOS are to: (i) design a complex SOS Type 2 architecture, and (ii) develop MS&A models to characterize and predict the behavior of the complex SOS Type 2 architecture. Like SoS perspective, in SOS, most if not all individual systems and systems' components are already in place. Therefore, the development of a complex SOS Type 2 architecture requires a full consideration of the technical feasibility of the constituent systems within Family of Systems (FoS), systems within different types of FoS, and the compatibility between the SOS requirements (i.e., systems of different FoS' systems) and those of individual systems' components. Practically, the complex SOS architectural focus could be just to establish adequate communication protocols between systems and ensure proper interoperability. If the SOS requirements is distinctively unique, a more elaborate SOS architecture design is required to modify some existing constituent systems to achieve overall SOS effectiveness. As such, the task of SOS architecture design and analysis is to strike a balance between the enterprise goals of all involved stakeholders. Reference [14] employs this SoS requirement-based approach to address architecture design challenges from mission definition to architecting complex SOS. However, for complex SOS Type 2 architecture with different FoS types, with requirements that are not stable due to the technology changes and dynamic operational environment conditions for one type of FoS as compared to the other types of FoS, the requirement-based design approach may lead to a stove-pipe architecture solution due to the following reasons:

- As mentioned above, requirements are usually derived based on the assumptions of selected technology enablers that are usually a few years behind the current technologies (e.g., at least 3 to 5 years for space systems),

- When the requirements of one type of FoS are changing at a fast pace, the As-Is system architecture design using SOS requirement-based approach becomes a "bottleneck" when interfacing with newer systems that are not within the same family,

- The family of SoS that requires frequent upgrade/refresh due to technology changes can cause a loss of interoperability with family of SoS that have stable requirements,

- Practically, SOS management at the enterprise level can pose a real challenge, when a family of SoS systems with stable requirements are not able to synchronize with the other families of SoS systems to be deployed at a later date; Managing this misalignment of systems and requirements synchronization is unmanageable task.

As discussed earlier, the current trends are to extend SoS requirement-based engineering to SOS capability-based engineering approach for the design and development of MS&A of complex systems. These current trends are based on the following capability-based engineering concepts:

- SOS architecture design is based on the top-down approach by associating each system with its high-level capabilities not subsystem's components' requirements. This approach leaves the flow-down of capability-to-requirement to the selected contractors for the design and build of each FoS,

- Managing the capabilities at the SOS enterprise level, i.e., SOS capability alignment, synchronization and integration will be managed using capability-based engineering approach.

- Develop MS&A models to simulate, analyze and characterize SOS capabilities with well-defined SOS architecture performance metrics focusing on capabilities, and not detail requirements' metrics.

4. Existing SOS Modeling, simulation and analysis (MS&A)

The current trends for the development of SOS MS&A models requires to classify and decompose complex SOS according to their presumed capability's attributes and relationships. In the context of this chapter, the classification and decomposition will be performed using SOS capability-based engineering perspective, which is also referred to as SOS perspective that encapsulate the three SOS types as defined earlier. Using [15], the book chapters have presented SOS taxonomy for space and airborne systems.

Optimizing the synergy between independent and heterogeneous systems or FoS' systems is a particular challenge in SOS architecture analysis. In order to analyze the SOS architecture effectiveness and to specify SOS characteristics, behaviors and features, engineers develop SOS MS&A frameworks and models and associated SOS performance metrics. The book chapters successfully demonstrated the use of SOS perspective for the development of MS&A models to characterize the performance of a notional space systems-of-systems with different FoS types.

5. A SOS perspective and current trends on the MS&A of decision support systems (DSS)

Decision-making is being profoundly challenged by the digitization of the business world and the rise of environmental uncertainty and risks. Augmented by digital technologies, decision makers have in their hands massive amount of open source data, and often, they are forced to make swift decisions while trying to mitigate increasing level of risks. Given the diversification of massive information sources that go beyond the organization boundaries, decision makers are facing with a triple level of uncertainty – increased difficulty in identifying the possible courses of actions given a complex decision; increased difficulty in estimating the likelihood of decision outcomes of a chosen action, and unexpected emergence of new actors – be it allies or foes – along the decision-making process.

DSS are commonly known as computer-based systems that are designed for aiding decision-makers to transform an ill-defined and unstructured problem into a well-defined and structured problem. In that process, the problem can be iteratively analyzed, and possible decision outcomes can be visualized. The role and function of a DSS is to guide the user throughout the decision-making process that eventually lead to a final decision – either optimal or satisfactory. In this DSS perspective, MS&A is seen as a set of tools that help decision makers to (i) express their preferences and needs, (ii) explore all possible solutions, (iii) perform evaluation analyses, and (iv) select the best possible preferred outcomes. In complex decision-making situations, a DSS is a computer-based system that supports its user(s) to make effective decisions in ill-structured problems.

A DSS is typically composed of four components [4]:

- A Data Component: that allows the decision maker to retrieve relevant data from all possible relevant sources to generate new alternatives and new perspectives in decision making;
- A Model Component: that enables selection, creation, and manipulation of models to help decision makers best capture the dimensions of the decision problems;
- An Interface Component: that provides context-sensitive and personalized interaction with the users; and
- A Communication Component: that facilitates structured and free-format information exchanges between users.

In a SOS environment, the level of sophistication of these four components increases with the size of interconnected systems. The overarching SOS engineering challenge is to deal with the uncertainty related to the design of interoperability and coordination. It is critical to design context-dependent interfaces between individual systems using technologies such as the Unified Modeling Language (UML), with the need for integration and coordination:

- SOS Dynamic Program Solver: Given a specific complex problem at hand, the DSS facilitates the generating of problem-solving solutions. It can simply be the execution of an existing ready-to-be-applied decision model, or it can set up a sequence of trial-and-error algorithms in search for an optimal problem formulation.
- SOS Daemon: In a highly distributed system, the role of this kernel of the multi-tasking DSS is to manage, coordinate and control in the background, without the direct control of the user, the overall execution of the SoS. Upon request, the daemon activates the processing of data, model and analysis, interpretation and visualization.
- SOS Scheduler: In a large-scale, loosely coupled distributed architectures, the SoS scheduler task is to generate a feasible scheduling scheme to execute decision making problems. It is providing process-driven policies to help the SOS Deamon to sequence DSS execution – either in a concurrent or sequential manner – and to allocate computational resources to each of the DSS tasks.
- SOS Planner and Controller: In highly vulnerable operations, the Planner and Controller performs the function of a hardware/software watchdog to ensure that the SoS can be promptly reset if it is disrupted by malfunctions or failures. In normal operations, the DSS should allow for the coordination of modeling and simulation processes.

Over the last thirty years, with the emergence of Decision Support Technology, decision makers have benefited from dedicated DSS applications, from engineering to business and healthcare to engineering. Elevating the role and function of DSS to support complex SoS presents both opportunities and challenges to the discipline. As exemplified in the book chapters, the concept of DSS can offer group decision for risk management and collective generation of social urban policies. Furthermore, game theories can be used to engage antagonists in finding a solution that would be acceptable to all.

6. A SOS perspective and current trends on MS&A of game theoretical models (GTM)

The development of computer models to perform the MS&A of a particular game theoretical model, whether it is used for analyzing and predicting the behaviors of complex systems or future economic well-being or forecasting stock market, are very challenging for system analysts, economic analysts, financial analysts and mathematicians. As pointed out in the book chapters, practical GTM are usually very complex because these models require the analysts and mathematicians to have a deep understanding of a combination of:

(i) Game theory, i.e., what type of game that the analyst should be selected for the problem being investigated and how-to set-up the game to achieve desire outcomes,

(ii) Advanced mathematics[3] in several fields, including modeling and simulation, algebra, probability, statistics and calculus, and

(iii) The problems being investigated with profound subject matter expertise in a particular scientific area. As an example, the analyst may require having a thorough understanding of the behavior of the cellular system being investigated or understanding of several medical fields, including tumor cells, thoracic surgery, neocortical epilepsy and kidney donation.

Currently, researchers, scientists, engineers and analysts have applied game theory to many applications, including but not limited to health care, wireless communications systems, complex systems acquisition, finances, and economics. As mentioned in Section 4 above, for the development of SOS MS&A models using SOS perspective, a game theoretical modeling taxonomy is required. **Figure 1** provides a description of a game theoretical taxonomy that the SOS modelers can use in selecting proper game "engine" (or system in SOS context) to perform required SOS MS&A modeling tasks. As shown in **Figure 1**, there are two types of games, namely, static games and dynamic games. For each game type, depending on the information available to the players, the MS&A modelers can choose to play complete or incomplete information with either corporative or non-corporative games. For static games, the modelers develop game engines using normal form. While, dynamic games use extensive form with the complete information games being divided into perfect information and imperfect information games. The definitions for these games are given below:

- Complete information games: Players know all information about the other players, e.g., their "types", strategies, payoffs and preferences.

- Incomplete information games: Players may or may not know some information about the other players, e.g., their "types", strategies, payoffs or their preferences.

- Perfect information games: Players are aware of the actions chosen by other players. They know who the other players are, what their possible strategies/actions are, and the preferences/payoffs of these other players. Information about the other players in perfect information game is also complete.

[3] Note that mathematical game theory can also be considered as a brand of mathematics that required mathematical background in modeling and simulation, algebra, probability, statistics and calculus.

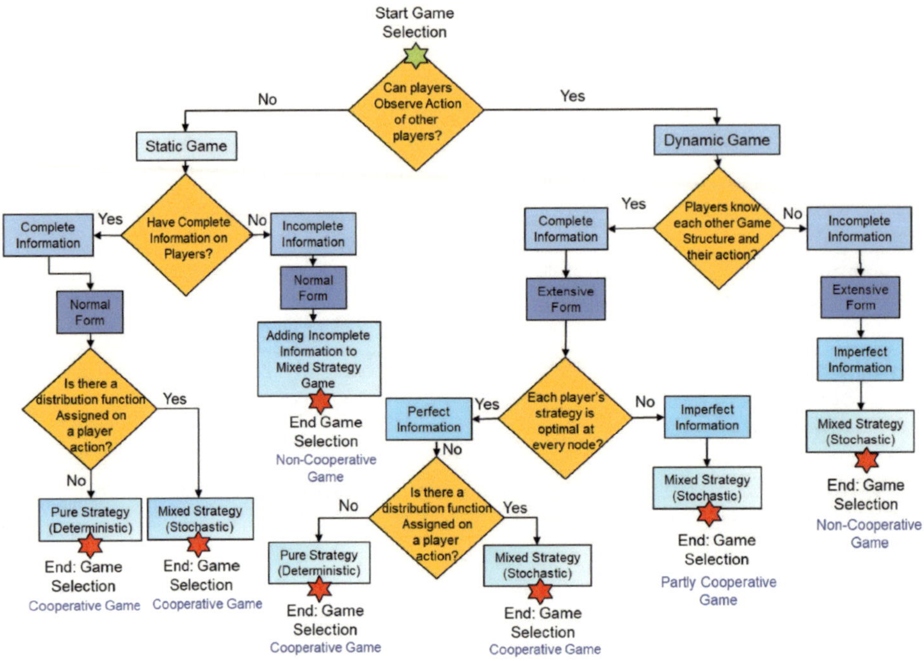

Figure 1.
Game theoretical Modeling Taxonomy from SOS perspective.

- Imperfect information games: Players simply do not know of the actions chosen by other players. However, they know who the other players are, what their possible strategies/actions are, and the preferences/payoffs of these other players. Hence, information about the other players in imperfect information game is incomplete.

Figure 1 assumes all games are Bayesian games. Basically, there are seven game models associated with static and dynamic game types. Each of these game models can be considered as a model of a "system" with a specified set of inputs and outputs and when these systems are set-up properly, together they will provide desired outcomes. Recently, References [16, 17] have used this SOS MS&A approach for developing complex computer simulation models to evaluate and develop acquisition strategy for acquiring complex space systems for U.S. DoD. Reference [18] discusses the use of decision support system to complement the game models when decisions on a selecting of a particular architecture solution does not converge. Therefore, it is probably safe to claim that the current trends for the development of complex MS&A models for game theoretical modeling applications are to use (i) SOS perspective, and (ii) Decision support system when the game model of a complex system or process does not converge.

7. Conclusion

The chapters presented in this technical book share a common thread of using Systems-of-Systems (SOS) perspectives and existing INCOSE System-of-Systems (SoS) engineering to address complex practical test-and-evaluation processes and systems, ranging from melt spinning process and space systems enterprise to medical problems. While dealing with a great level of details illustrating their thematic methodologies, the authors do take the effort in providing the readers with a thorough background discussion, making their chapters comprehensive and reachable to all. We hope that the readers will find the chapters selected for this book a source of ideas for their own work -- ideas that are anchored in a set of inter-disciplinary and, moreover, ideas are being explored or experimented in promising systems-of-systems applications.

By its very nature, an SOS evolves overtime with changing systems requirements. Looking forward, the coordination of simultaneous and real-time processes in large-scale programmable networks remains to us a major challenge in the design of SOS for complex problem solving. In spite of the recent advances in Artificial Intelligent (AI) and automated systems, the burden of managing the interfaces between distributed data repositories, disparate and incompatible model bases, and context-dependent visualization and presentation techniques, the coordination of parallel operations still depends in a large part on the SOS designers. It is a time-consuming and tedious process to conceptualize and implementing parallelism. We hope the readers find in this book concepts and ideas that would help them achieve their requirements analysis and SOS design task.

Last but not least, and although not explicitly discussed, most of the chapters in this book have judiciously pointed out the necessity of establishing a coherent set of principles for systems-of-systems engineering, including synergism, symbiosis, modularity and self-governance, conservation and reconfiguration, emergence and re-architecting, efficiency and effectiveness. If these principles are clearly defined and adopted by the community of SOS developers and users, we expect that SOS would be the next and ultimate generation of digital applications.

Author details

Tien M. Nguyen[1,2*] and Tung X. Bui[3,4,5]

1 Center for Computational and Applied Mathematics (CCAM), California State University, Fullerton (CSUF), USA

2 The Aerospace Corporation, USA

3 Shidler College of Business, USA

4 Information Technology Management, University of Hawaii at Manoa, USA

5 Haas School of Business, University of California at Berkeley, USA

*Address all correspondence to: tmnguyen57@fullerton.edu

References

[1] U.S. DOD, Systems Engineering Guide for Systems of Systems, Version 1.0, August 2008. Available at: https://acqnotes.com/wp-content/uploads/2014/09/DoD-Systems-Engineering-Guide-for-Systems-of-Systems-Aug-2008.pdf

[2] R.W. Asby, Introduction to Cybernetics. London: Routledge Kegan & Paul, 1964.

[3] L. Von Bertalanffy, General System Theory: Foundations, Development, Applications. New York: George Braziller, 1968.

[4] Tung X. Bui, Co-oP: A Group Decision Support System for Cooperative Multiple Criteria Group Decision Making. Berlin: Springer, 1987.

[5] H. Raiffa, Decision Analysis: Introductory Lectures on Choices under Uncertainty. Massachusetts: Addison-Wesley, (968.

[6] H. Simon, The Sciences of the Artificial. Massachusetts: MIT Press, 1969.

[7] R. Schlaifer, Analysis of Decisions under Uncertainty. New York: McGraw-Hill, 1969.

[8] J. Von Neumann and O. Morgenstern, Theory of Games and Economic Behavior. New Jersey: Princeton University Press, 1953.

[9] J.F. Nash, The Bargaining Problem. In Econometrica. Econometric Society. 18(2): 155-162, 1950.

[10] H.F. Von Stackelberg Market Structure and Equilibrium, 1934. 1st Edition (translation into Englsih), Bazin, Urch & Hill: Springer (2011).

[11] H. Kopetz, System of Systems Challenges. In: Schoitsch E. (eds) Computer Safety, Reliability, and Security. SAFECOMP 2010. Lecture Notes in Computer Science, vol 6351. Springer, Berlin, Heidelberg. https://doi.org/10.1007/978-3-642-15651-9_35.

[12] A.M. Pendleton-Julian and J.S. Brown, Design Unbound. Massachusetts: MIT Press, 2018.

[13] INCOSE, New Standards for System-of-Systems (SoS) Engineering. Available at: https://conference. conflr.com/events/Demo/showcases/incose/INCOSE%20System%20of%20Systems%20Engineering%20Standards.pdf.

[14] Imane Cherfa, Nicolas Belloir, Salah Sadou, Régis Fleurquin and Djamal Bennouar, "Systems of Systems: From Mission Definition to Architecture Description," Wiley Systems Engineering. 2019; 1-18. Available at: https://wileyonlinelibrary.com/journal/sys.

[15] J. Gideon et al., "Taxonomy of Systems-of-Systems," Conference on Systems Engineering Research, Institute of Electrical and Electronics Engineers (IEEE), Jan 2005.

[16] Tien M. Nguyen, Andy T. Guillen, Sumner S. Matsunaga, Hien T. Tran and Tung X. Bui, War-Gaming Applications for Achieving Optimum Acquisition of Future Space Systems, Book Chapter in Simulation and Gaming Book, 2018, IntechOpen Publisher. Available at: https://www.intechopen.com /books/simulation-and-gaming/war-gaming-applications-for-achieving-optimum-acquisition-of-future-space-systems.

[17] Tien M. Nguyen, Hien T. Tran, Andy T. Guillen, Tung X. Bui, and Sumner S. Matsunaga, "Acquisition War-Gaming Technique for Acquiring Future Complex Systems: Modeling and Simulation Results for Cost Plus

Incentive Fee Contract," Journal of Mathematics, 2018. Available at: https://doi.org/10.3390/math6030043.

[18] Tien M. Nguyen, Tom Freeze, Tung Bui, Andy Guillen, "Multi-Criteria Decision Theory for Enterprise Architecture Risk assessment: Theory, Modeling and Results," 2020 SPIE Conference Proceedings, Conference: Sensors and Systems for Space Applications XIII; DOI:10.1117/12.2559317.

Section 2

Systems-of-Systems (SOS) Engineering

Chapter 2

Systems-of-Systems Taxonomy: Space and Airborne Systems Perspective

John D. Nguyen

Abstract

This chapter discusses the taxonomy of Systems-of-Systems (SOS) with a focus on space and airborne systems perspective. A discussion with a broad view of taxonomy with considerations for space and airborne systems classification, including SOS and Family-of-Systems (FOS), will be presented. The chapter defines taxonomic categories considering dimensions in the classification of space and airborne SOS based on their acquisition strategy, operational mode, and problem domain. Commercial and military acquisition strategies will be addressed along with their intentional operational modes and problem domains. The space and airborne systems discussed will be Satellite Communication (SATCOM) systems, sensing and imaging satellite systems and Positioning-Navigation-and-Timing (PNT) satellite, and military and commercial aircraft systems. The chapter provides examples on notional military SATCOM and manned aircraft systems.

Keywords: space systems, systems-of-systems (SOS), satellite communication, sensing and imaging satellite, positioning-navigation-and-timing (PNT) satellite, family of system (FOS), SOS engineering, constituent systems, acquisition, autonomy of constituents, application domains, standards, operational Independence, managerial Independence, evolutionary development

1. Background and introduction

The term "Taxonomy" used in this chapter will be borrowed from the definitions presented in Refs. [1–3], but with an emphasis on space and airborne perspective. This chapter defines taxonomy as a hierarchical structure framework to classify space and airborne systems terms into parent-and-child relationships, where each level of a hierarchy can be referred to as a "Category." In this chapter, "Systems" will be categorized as Systems of Systems (SOS) and Family of Systems (FOS). For general military space systems, military space FOS can be categorized as (i) Satellite Communication (SATCOM) systems, (ii) Sensing and Imaging satellite systems, and (iii) Positioning-Navigation-and-Timing (PNT) satellites. Practically, civilian space FOS can be categorized as (i) commercial FOS of Broadcasting satellites, (ii) commercial FOS of Wideband Internet satellites, and (iii) commercial FOS of Data, Video, Audio Communications satellites. In general, commercial space FOS can be categorized as (i) NASA FOS of Near-Earth Missions, (ii) NASA FOS Deep Space missions, and (iii) NOAA FOS Earth Surveillance satellites. SOS can be a selected group of FOS that are connected together. For space systems, they (SOS) can be categorized as military, civilian and commercial space systems. For airborne systems, they can be categorized as military and commercial airborne systems, since civilian and commercial airborne systems are practically identical. In general, airborne systems can be categorized as (i) military manned aircraft systems, and (ii) military un-manned aircraft systems. Similarly, for commercial aircraft, it is also can be categorized as (i) commercial manned aircraft systems, and (ii) commercial un-manned aircraft systems. **Figure 1** presents our view of taxonomy for "systems". **Figure 2(a)** and **(b)** illustrate taxonomies for Space and Airborne FOS, respectively.

This chapter will focus on SOS taxonomy for space and airborne systems. Due to the constraint on the length, this chapter will only provide two taxonomy examples on space and airborne systems, where a generic space and airborne systems' taxonomies will be presented. For airborne systems, the example will be focused on manned airborne systems. The chapter is organized as follows:

i. Section 2 describes existing SOS taxonomy framework described in [3];

ii. Section 3 discusses our proposed SOS taxonomy framework for space systems;

iii. Section 4 proposes an approach for SOS taxonomy framework for airborne systems;

iv. Section 5 provides examples on notional military SATCOM and manned aircraft systems;

v. Section 6 concludes the chapter with remarks on the taxonomy's needs for future space systems.

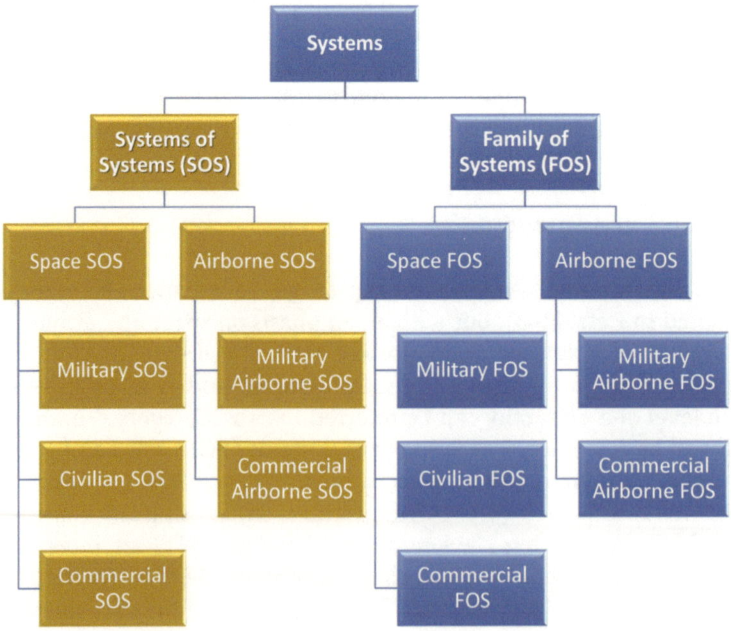

Figure 1.
A taxonomy definition of systems.

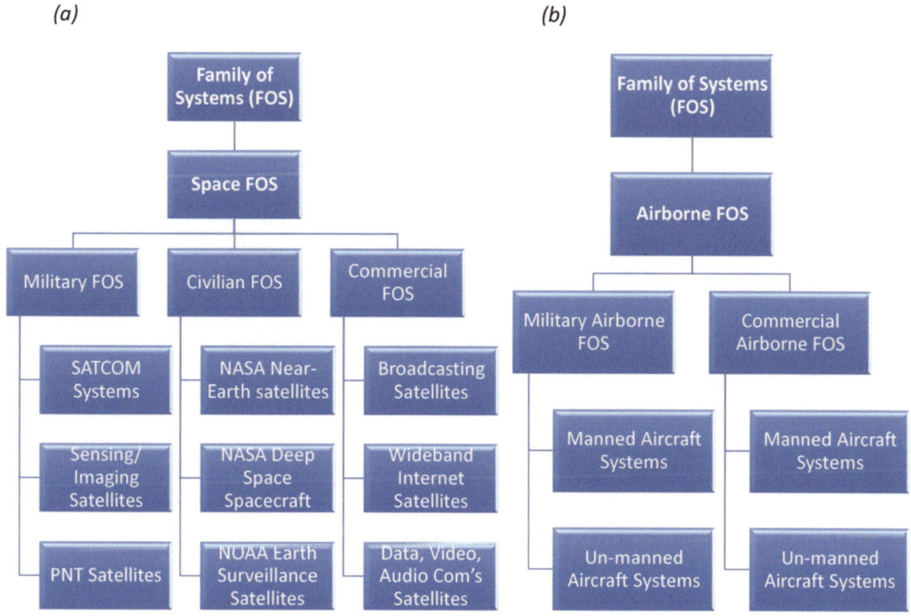

Figure 2.
Taxonomy definitions of space and airborne FOS (a) a taxonomy definition of space FOS (b) a taxonomy definition of airborne FOS.

2. Description of existing SOS taxonomy framework

This section describes a current perspective on the SOS taxonomy framework, where SOS can be categorized as three types:

- Acquisition type
- Operation type
- Domain type.

Each of the type will be classified into different component as shown in **Figure 3**. The following provides a summary of the three types and their associated components.

- **Acquisition type:** SOS is classified based on how the systems acquired [3]:
    - **Dedicated SOS:** Is defined as planned SOS, where they are consciously designed and engineered from the beginning to be SOS, where the interaction between the component systems is expected when the systems are acquired. As pointed out in [3], in the past, many military SOS were not acquired in this manner, and the emerging trend is to design military systems around the SOS concept.
    - **Virtual SOS:** Unlike dedicated SOS, this type of SOS is un-planned when the component systems are engineered and acquired. Another characteristic of these systems is that once their use has ended the

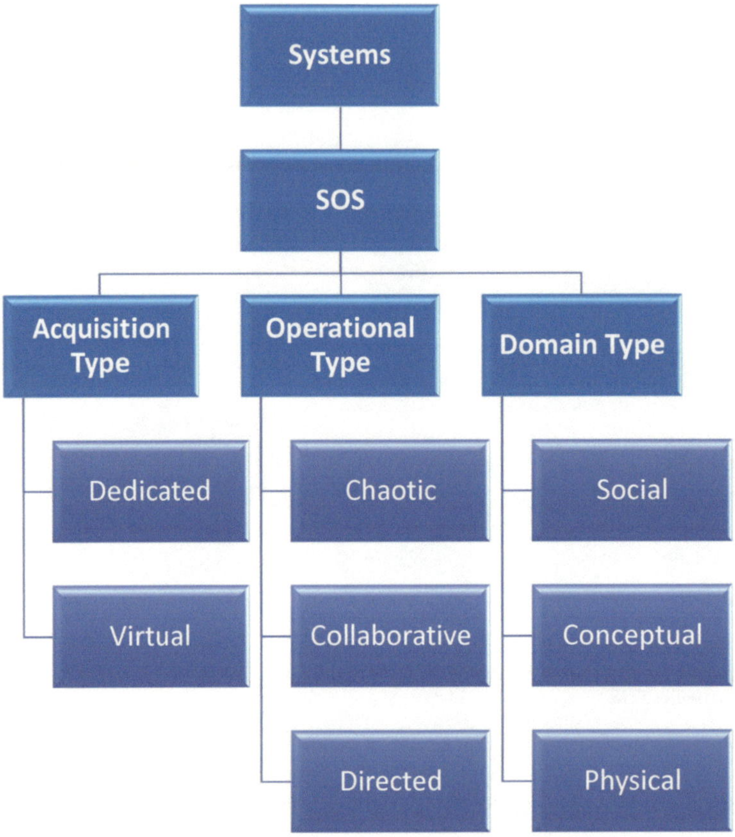

Figure 3.
Current SOS taxonomy framework [3].

component systems are usually disassembled and no longer operate as a part of a larger SOS.

- **Operational type:** SOS is classified based on how the systems operate [3]:

    - **Chaotic SOS:** This type of SOS has no central control authority or managerial entity and thus no agreed upon purpose.

    - **Collaborative SOS:** For this type of SOS, the component systems interact voluntarily almost out of necessity. Control and management authority have little power to coerce the behavior of the component systems. Control and Management authorities may issue standard practices and procedures by which components must operate to be a part of the larger system, but ultimately it is up to the component systems to acquiesce to those standards to become part of the larger system (as with the Internet). The overall behavior of these systems may still be somewhat unpredictable.

    - **Directed SOS:** This type of SOS is designed to have its control by a central management authority. The systems are designed and operated for a specific purpose.

- **Domain type:** SOS is classified based on how the domain that systems operate [3]:

- **Social SOS:** Is defined as SOS that are either physical or conceptual SOS classes. As pointed out in [3], a social SOS is government.

- **Conceptual SOS:** These systems do not exist as tangible entities in physical space nor do they operate on or manipulate matter. Systems that are conceptual include those in which humans interact with concepts or those that require no human intervention at all.

- **Physical SOS:** These systems are operating in or on the physical world. These systems involve interactions between humans and the physical world or systems that are completely embedded in the physical world with no human interaction. These systems are composed of component systems that are tangible or affect matter.

3. Proposed SOS taxonomy framework for space systems

Figure 4 describes the proposed SOS taxonomy for space systems, where SOS can be categorized as military SOS, civilian SOS and commercial SOS. As discussed in Section 1, military SOS can be classified SATCOM SOS, Image/Sensor SOS, PNT SOS, and mixed SATCOM + Image/Sensor + PNT SOS. Similarly, the classifications for civilian and commercial SOS are also shown in **Figure 4**.

Each type of the military SOS (e.g., SATCOM SOS) can be further classified in terms of acquisition type, operational type and domain type. Using the current framework described in the above section [3], this section derives the proposed taxonomy framework for space systems. The section is organized as follows: (i) Section 3.1 presents a taxonomy framework for military space systems,

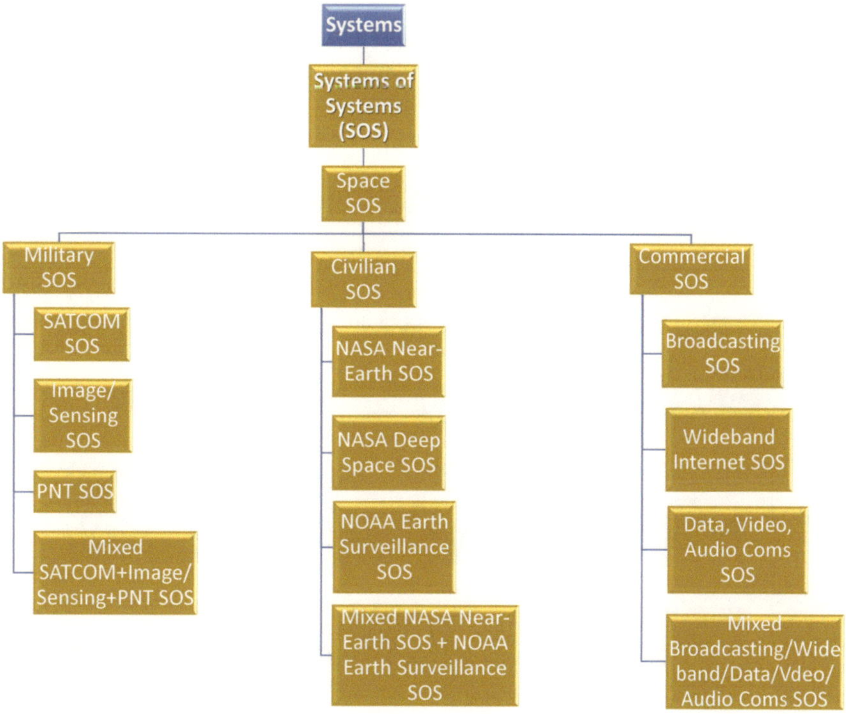

Figure 4.
Proposed SOS taxonomy framework for space systems.

(ii) Section 3.2 addresses the commercial space systems, and (iii) Section 3.3 discusses the taxonomy framework for civilian spaces systems.

3.1 SOS taxonomy framework for military space systems

Figure 5 presents our proposed SOS taxonomy for military space systems. The SOS taxonomy framework for military space systems can be categorized as:

- Acquisition type
    - Dedicated SOS
    - Virtual SOS
- Operation type
    - Collaborative SOS
    - Directed SOS
- Domain type:
    - Physical SOS

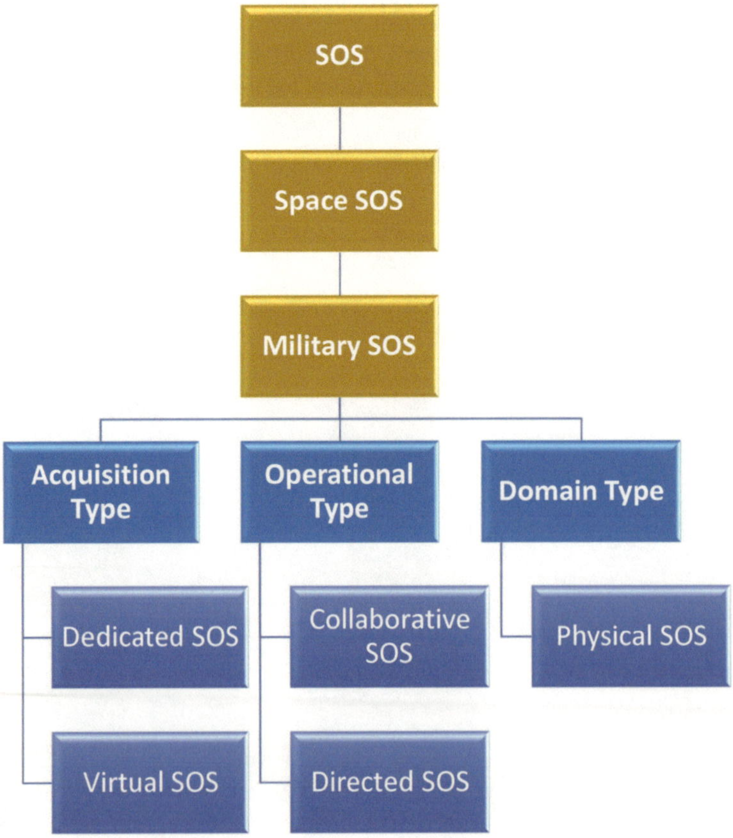

Figure 5.
Proposed SOS taxonomy framework for military space systems.

3.2 SOS taxonomy framework for commercial space systems

Figure 6 presents our proposed SOS taxonomy for commercial space systems. The SOS taxonomy framework for commercial space systems can be categorized as:

- Acquisition type
  - Dedicated SOS
- Operation type
  - Collaborative SOS
  - Directed SOS
- Domain type
  - Physical SOS

Note that the commercial framework missed a component in the acquisition type, namely, virtual SOS. This is because the SOS solution is usually derived from customer's needs. While for military space systems, due to the threats dynamic, the warfighter needs are changing at a fast pace and there will be un-planned SOS components to be deployed.

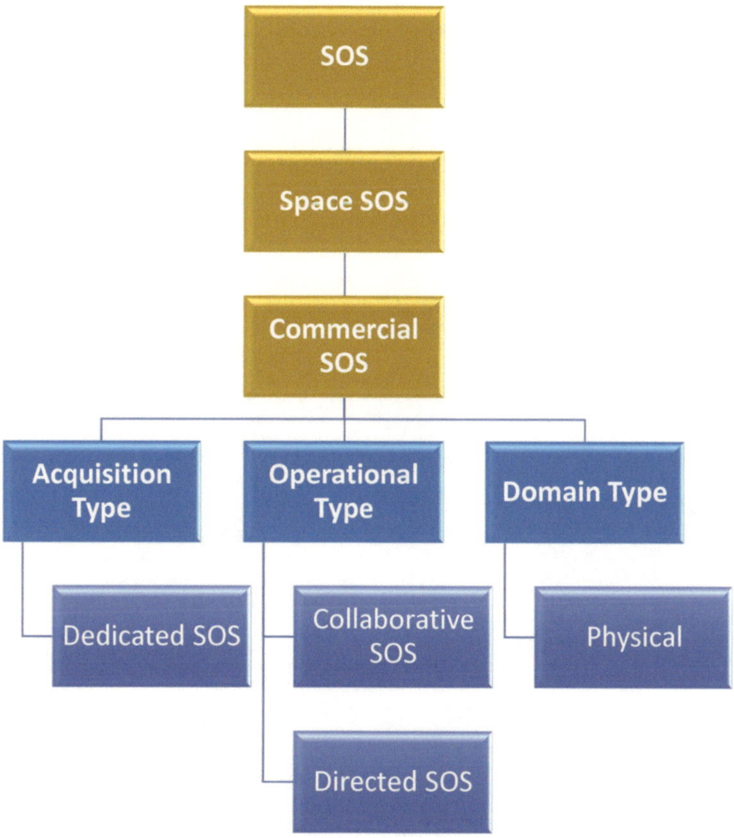

Figure 6.
Proposed SOS taxonomy framework for commercial space systems.

3.3 SOS taxonomy framework for civilian space systems

Figure 6 presents our proposed SOS taxonomy for commercial space systems. The SOS taxonomy framework for civilian space systems can be categorized as:

- Acquisition type
    - Dedicated SOS
- Operation type
    - Collaborative SOS
    - Directed SOS
- Domain type
    - Physical SOS

Note the proposed SOS taxonomy framework for civilian space systems is identical to commercial space systems, see **Figure** 7. Similar to commercial space systems, the SOS acquisition for civilian space systems is depending on a planned mission's needs.

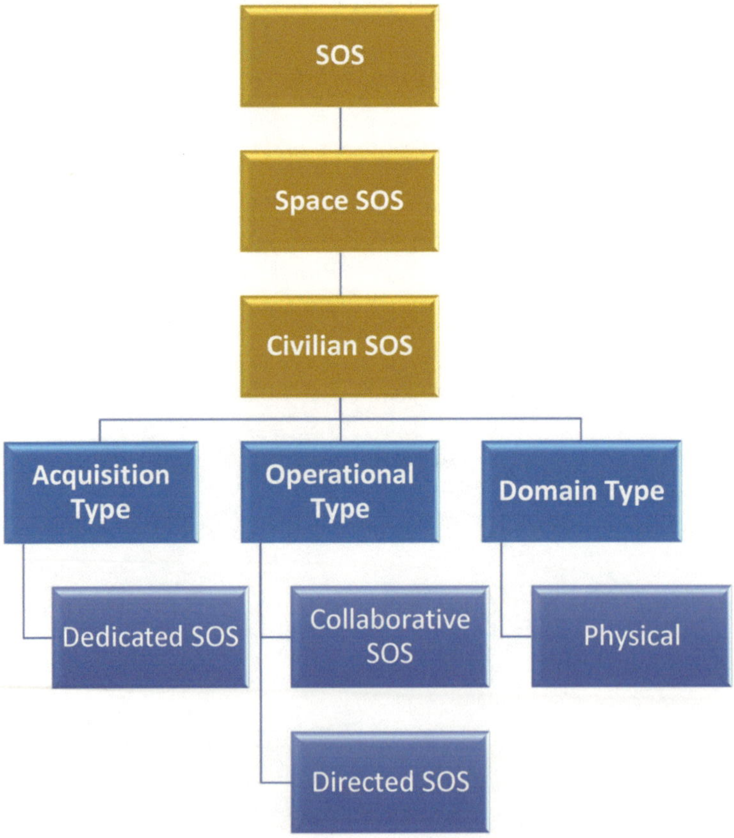

Figure 7.
Proposed SOS taxonomy framework for civilian space systems.

4. Proposed SOS taxonomy framework for airborne systems

Again, using the framework described in Section 2, this section derives the proposed taxonomy framework for airborne systems. The section is organized as follows: (i) Section 4.1 presents a taxonomy framework for military airborne systems, and (ii) Section 4.2 addresses the commercial space systems. Note that the taxonomy framework for civilian airborne systems is identical to commercial systems.

4.1 SOS taxonomy framework for military airborne systems

Figure 8 presents our proposed SOS taxonomy for military airborne systems. The SOS taxonomy framework for military airborne systems can be categorized as:

- Acquisition type
    - Dedicated SOS
    - Virtual SOS
- Operation type
    - Collaborative SOS

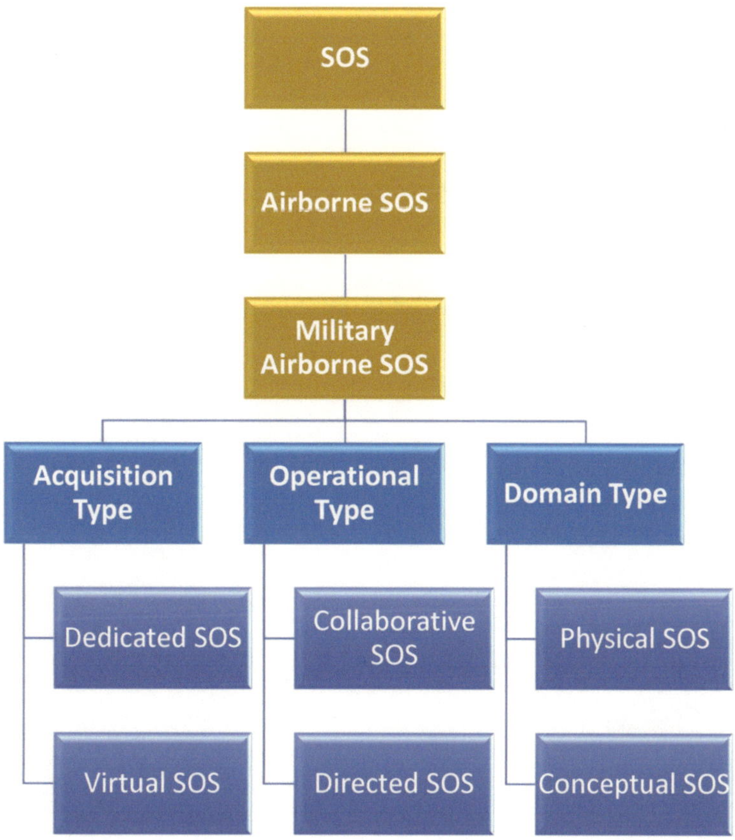

Figure 8.
Proposed SOS taxonomy framework for military airborne systems.

- Directed SOS

- Domain type

  - Physical SOS

  - Conceptual SOS

Similar to military space systems, due to the threats dynamic, the warfighter needs are changing at a fast pace and there will be un-planned SOS components to be deployed in airborne platforms. However, there is an addition component that is included in the domain type, namely, conceptual SOS. For airborne systems, this conceptual SOS component provide pilot training systems that have components existed in both physical and non-physical domains. For the physical domain, the pilot training system includes the training facility. For the non-physical domain, the pilot training system includes the cyber space component, where the pilots encounter the non-physical entities for training purposes.

4.2 SOS taxonomy framework for commercial airborne systems

The SOS taxonomy framework for commercial airborne systems is very similar to the military airborne systems, except that the SOS component for acquiring un-planned systems is no longer required. Thus, the framework includes:

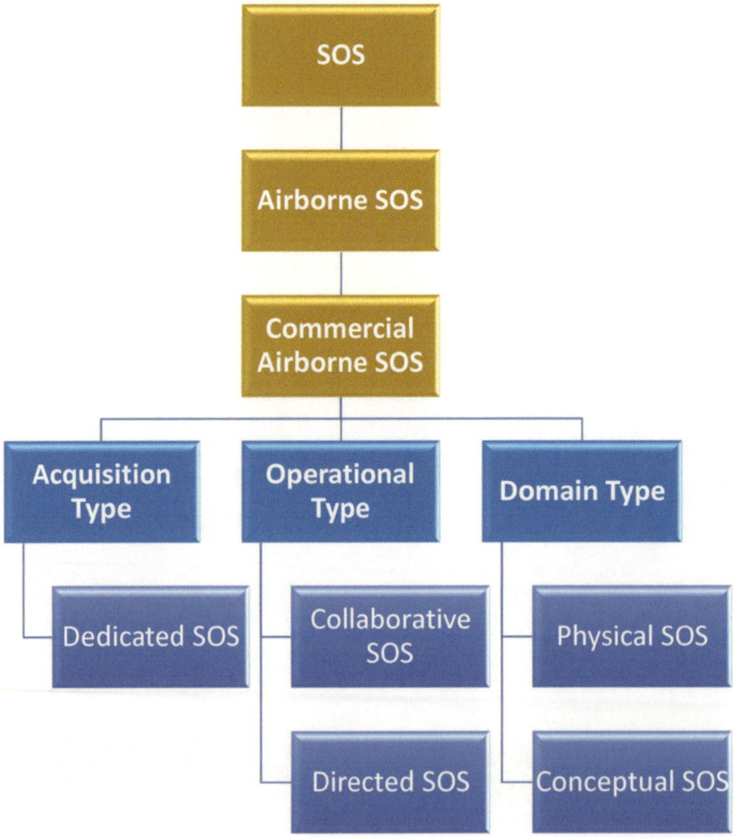

Figure 9.
Proposed SOS taxonomy framework for commercial airborne systems.

- Acquisition type
    - Dedicated SOS
- Operation type
    - Collaborative SOS
    - Directed SOS
- Domain type
    - Physical SOS
    - Conceptual SOS

Figure 9 depicts the proposed SOS taxonomy framework for commercial airborne systems.

5. Examples on notional military space and airborne systems

This section provides examples of taxonomy on notional military space and airborne systems since the case for civilian and commercial systems can also be derived directly from these examples. **Figure 10** illustrates the two examples to be described in the following subsections. Subsection 5.1 presents an example of a taxonomy framework for a typical military SATCOM SOS. Subsection 5.2 provides an example of a taxonomy framework for a typical manned military aircraft that

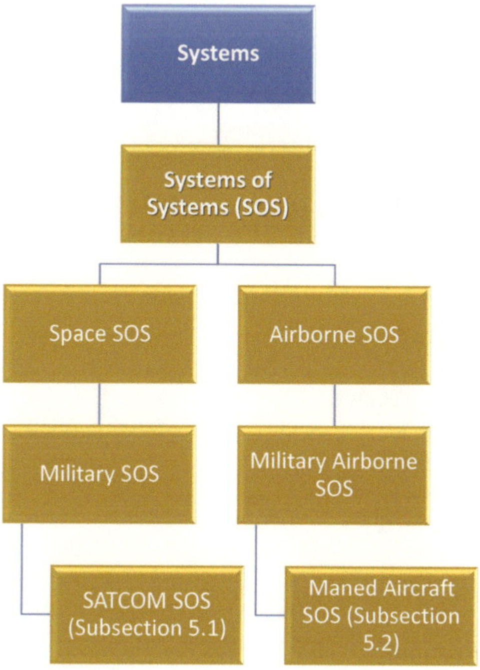

Figure 10.
SOS taxonomy framework for military space and airborne systems (described in the following subsections).

can be used for the derivation of a taxonomy framework for un-manned military aircraft.

5.1 A notional military SATCOM system

A typical military SATCOM system with a fully process SATCOM payload can be classified into the following subsystem components that are part of a SOS design [4]:

- Received Antenna Subsystem (RX-AS)
- RF Front-End Subsystem (RF-FES)
- Fully Digital Processing Subsystem (FDPS)
- Tracking-Telemetry & Commanding Subsystem (TT&CS)
- Frequency & Timing Subsystem (FTS)
- Altitude & Control Subsystem (A&CS)
- Communication Security Subsystem (COMSECS)
- RF Back-End Subsystem (RF-BES)
- Transmit Antenna Subsystem (TX-AS)

Followings are the decomposition of each of the above subsystems. Typical RX and TX antenna subsystems include the following components [4]:

- Antenna Configurations
- Beamformer Component
- Antenna Controller Component.

Typical RF Front-/Back-End Subsystems include the following components [4]:

- Front-end
    - Low Noise Antenna (LNA)
    - Multi-wideband Receiver
    - Down RF Converter
    - Tunable IF Converter
- Back-end
    - High Power Amplifier (HPA)
    - Up RF Converter

The FDPS is the heart of a fully processing payload, and a typical FDPS includes the following components [4]:

- Analog-to-Digital Converter/Digital-to-Analog Converter (ADC/DAC)
- Digital Processor (e.g., Field Programmable Gate Array (FPGA))
- Modulator/Demodulator (MOD/DEMOD)
- Decoder/Encoder
- Digital Network Switch (DNS)
- Fully On-Board Satellite System Controller (FOBSSC)

A typical TT&CS includes the following components [4]:

- On-Board Ranging Processor
- On-Board Command Processor
- On-Board Telemetry Processor

Using a typical fully process SATCOM payload described above, **Figure 11** provides a SOS taxonomy framework for a notional military SATCOM system operating in an SOS environment.

5.2 A notional military airborne system

A typical military manned airborne system can be classified into the following subsystem components that are part of a SOS design [5]:

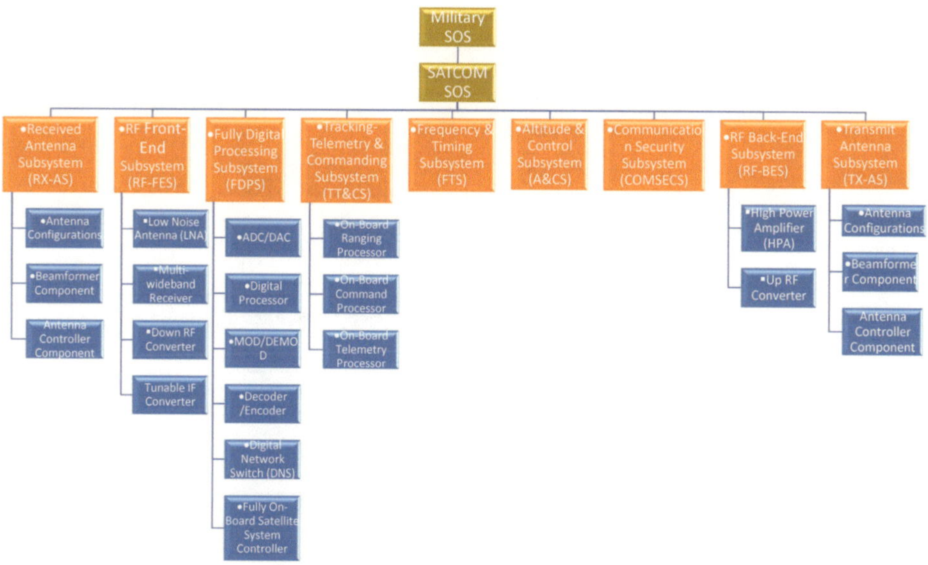

Figure 11.
SOS taxonomy framework for military SATCOM systems.

- Airframe
- Propulsion
- Application Software
- Communications / Friend-or-Foe (FoF) Identification
- Navigation / Guidance Control
- Central Computer

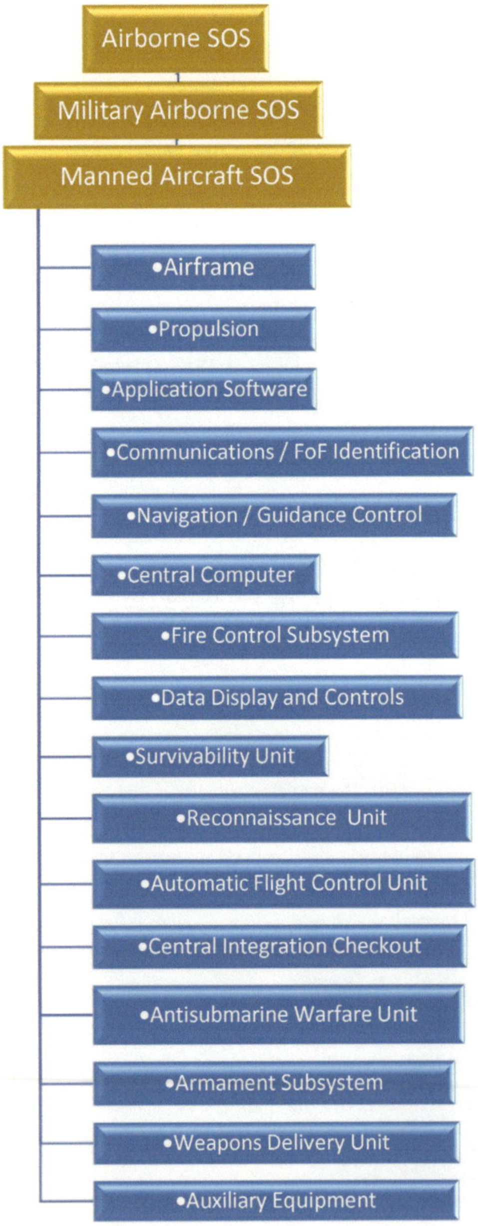

Figure 12.
SOS taxonomy framework for military manned aircraft.

- Fire Control Subsystem
- Data Display and Controls
- Survivability Unit
- Reconnaissance Unit
- Automatic Flight Control Unit
- Central Integration Checkout
- Antisubmarine Warfare Unit
- Armament Subsystem
- Weapons Delivery Unit
- Auxiliary Equipment

Figure 12 provides a SOS taxonomy framework for a notional military manned airborne system operating in an SOS environment.

6. Conclusion

This chapter provides descriptions of SOS taxonomy frameworks for space and airborne systems. A broad view of taxonomy frameworks with considerations for space and airborne SOS and FOS were presented. Using existing SOS taxonomy framework, the chapter proposed SOS taxonomy frameworks for space and airborne systems based on their acquisition types, operational modes, and problem domains. Examples on SOS taxonomy were provided for notional military SATCOM and manned aircraft systems. Similarly, examples for manned airborne and other space systems can also be derived by tailoring the framework presented in Section 5.

Conflict of interest

The preparation of this chapter was not funded by Gulfstream Corporation, and it was done by the author using his own time and resources, thus it does not represent Gulfstream's view on airborne system taxonomy.

Notes/Thanks/Other declarations

The author wishes to thank his wife, Annie Luu-Nguyen, for her immense patience and support.

Author details

John D. Nguyen[1,2]

1 John DTN Consulting Services, Huntington Beach, California, USA

2 Gulfstream Corporation, Long Beach, California, USA

*Address all correspondence to: johndncva@gmail.com

References

[1] Walli B. Taxonomy 101: The basics and getting started with taxonomies. Knowledge Management (KM) World. 15 August 2014. Available from: https://www.kmworld.com/Articles/Editorial/What-Is/Taxonomy-101-The-Basics-and-Getting-Started-with-Taxonomies-98787.aspx

[2] Taxonomy vs. Data. Available from: https://www.dpci.com/insights/taxonomy-vs-metada

[3] Gideon J, et al. Taxonomy of systems-of-systems. In: Conference on Systems Engineering Research, Institute of Electrical and Electronics Engineers (IEEE); 2005

[4] Nguyen JD. Overview of existing satellite systems and future trends of advanced satellite architectures. In: Nguyen TM, editor. Satellite Systems - Design, Modeling, Simulation and Analysis. Rijeka: IntechOpen Publisher; 2020

[5] MIL-HDBK-881, Department of Defense Handbook: Work Breakdown Structure (WBS) for Defense Materiel Items (Jan 2, 1998) [S/S BY MIL-STD-881C]

Chapter 3

Applications of Systems Engineering Technical Process Flows on Enterprise Systems

Alper Pahsa

Abstract

Systems engineering (SE) and SE management is the objective of all SE efforts, which defines the transformation of specific customer needs into a system product, service, or enterprise systems. Enterprise systems of systems engineering apply systems engineering fundamentals to the design of an enterprise. It is created by knowledge, principles, and processes tailored to the design of enterprise systems. Enterprise is a complex, socio-technical system that includes interdependent resources of people, information, and technology to reach a common goal. Enterprise systems is complex that a system configuration can be controlled among the different stakeholders. There are four different steps in enterprise systems process; it includes technology planning (TP), capabilities-based engineering analysis (CBEA), enterprise architecture, and enterprise analysis and assessment. This is the main reason that the enterprise work is developed and established at HAVELSAN Inc., Information and Security Technology Division. SE and technical processes for enterprise projects require establishing a systematic taxonomy and SE process customization. This chapter presents the work done on SE for enterprise projects at HAVELSAN. The chapter presents the results of the study of similarities and differences of the various applications of systems engineering of product systems oriented against enterprise systems.

Keywords: enterprise systems engineering, enterprise systems, system engineering (SE)

1. Introduction

SE approach is defined as [1]: "An interdisciplinary approach to evolve and verify an integrated and life cycle balanced set of systems product and process solutions that satisfy customer needs. Systems engineering: (a) encompasses the scientific and engineering efforts related to the development, manufacturing, verification, deployment, operations, support, and disposal of systems products and processes; (b) develops needed user training equipment, procedures, and data (c) establishes and maintains configuration management of the systems; (d) develops work breakdown structures and statements of work; and (e) provides information for management decision making."

The real-world application of SE is no different than its definition. In most practical applications, the knowledge about required product, service, enterprise, and system of systems (SoS) is mandatory for the successful project realizations.

The SoS is used in the literature since the 1950s to define systems that include independent constituent systems, which behave mutually with a common goal to establish a process. An example of SoS arises in the fields of power grid technology, transport, production, business, government, and military enterprises. SoS engineering links with independence, heterogeneity, evolution, and emergence properties of SoS. SoSs are found in wide areas. For instance, emergency response (fire, police, and hospital) with independent and managed systems never collaborates to provide a service on which trusted service is placed. Many conclusions can be drawn in **Figure 1** depending on the current state of SoSE.

Initially the literature is fragmented. Because SoSE is in the early stages, development process is preceded on the topic [2].

SoS for enterprise systems is a group of activities related to capability delivery design and establishment in enterprise mission planning. It transforms and realizes the enterprise goals (strategic policies) into an informative and consistent Enterprise Architecture as a Strategic Plan for "System of Systems" evolution of the Enterprise. It is a general principle that organization-level transformation needs are wide and detailed. They involve specific objective in mind of the enterprise requirements and transformation of those requirements in the enterprise. Since the enterprise is a complex system of systems, it needs a SoS enterprise systems engineering method with low-risk process to devise and manage such an organization-wide transformation [3].

The SoS ESE method includes a process to manage enterprise transformation throughout three critical components. Initially it needs comprehensive, integrated, and mission-service-based architecture. Secondly, it defines a management objective for using that architecture through all organization processes of the acquisition methods to minimize risk by linking the architecture's engineering processes to

Author	SYSTEM (S) OF SYSTEMS PERSPECTIVE
Sage and Cuppan (2001)	Systems of systems exist when there is a presence of a majority of the following five characteristics: operational and managerial independence, geographic distribution, emergent behavior, and evolutionary development. Primary focus: Evolutionary acquisition of complex adaptive systems. Application: Military.
Carlock and Fenton (2001)	Enterprise Systems of Systems Engineering focused on coupling traditional systems engineering activities with enterprise activities of strategic planning and investment analysis. Primary focus: Information intensive systems. Application: Private Enterprise.
Pei (2000)	System of Systems Integration to pursue development, integration, interoperability, and optimization of systems to enhance performance in future battlefield scenarios. Primary focus: Information intensive systems integration. Application: Military.
Lukasik (1998)	Integration of systems into systems of systems that ultimately contribute to evolution of the social infrastructure. Primary focus: Education of engineers to appreciate systems and interaction of systems. Application: Education.
Kotov (1997)	Systems of systems are large scale concurrent and distributed systems that are comprised of complex systems. Primary focus: Information systems. Application: Private Enterprise.
Manthorpe (1996)	In relation to joint warfighting system of systems is concerned with interoperability and synergism of Command, Control, Computers, Communications, and Information (C4I) and Intelligence, Surveillance, and Reconnaissance (ISR) Systems. Primary focus: Information superiority. Application: Military.

Figure 1.
Perspectives of systems of systems [2].

the acquisition processes. Lastly, by design it defines a flexible architecture intended to do leveraging industry best practices rather than forcing a prescriptive, one-size-fits-all of the process constraints [3].

The extent to which these combinations are selected from the previously determined approaches vs. the need for system engineers to create such combinations is part of SE application. This knowledge also lays the basis for the various application domains. Some domains have very detailed set of procedures, guidelines, and standards relevant to that domain, while others take general SE and tailor it as needed by using the judgment of those involved. In general, all domains have a little of both domain-specific guidelines and experienced people [4].

As discussed above, SE approaches transform innovations and engineering challenges into specific products, services, or enterprise systems. Based on these facts, HAVELSAN has been working on in a CMMI V3-based development product life cycle system and has been applying ISO/IEC 15288 SE standards traditional product development technical and engineering processes. Based on the structural differences in projected system needs, there exists a need to customize and develop new approaches of SE technical processes for the anticipated products and services [4] for enterprise system projects. For instance, in Information and Security Technologies Division at HAVELSAN Inc., there is more than 30% of projects that can be classified as service systems, more than 50% as product development projects, 4% as enterprise systems [5] project, and 16% as a combination of these systems. These facts require devising a new methodology of technical engineering processes for the new type of systems projects according to the ISO/IEC 15288 Systems and Software Engineering—System life cycle processes. In the following sections, fundamentals of product systems and enterprise systems are discussed, and then customized SE technical process for enterprise systems is compared against the product development SE and ISO/IEC 15288 Systems and Software Engineering technical processes. Finally, a conclusion will be provided.

2. Product systems

As defined in finance terminology, the word "product" was first defined as in economics and commerce products belonging to broader category of goods. In marketing, a product is anything that can be offered to a market that might satisfy a customer's needs. In retail industry, products are called merchandise. In manufacturing, products are purchased as raw materials and sold as finished goods. Commodities are usually raw materials, such as metals and agricultural products, but a commodity can also be anything widely available in the open market. In project management, products are the formal definitions of the project deliverables that constitute or contribute to deliver the objectives of the project. In insurance industry, the policies are considered products offered for sale by the insurance company that created the contract [2].

A product system is the combination of "end" products and the enabling products for those "end" products. This concept of a product is illustrated in **Figure 1** of Reference 5, the ANSI/EIA 632-2003 standard. The scope of the ANSI/EIA standard is clearly restricted to product systems described in **Figure 2** of [7].

The end product can be defined as a system with its own elements or subsystems, each of which has its own enabling products as stated in **Figure 3** [4]. The product development process usually focuses only on the SE of the end product. Performing SE processes for "Product Systems" is essential when either the

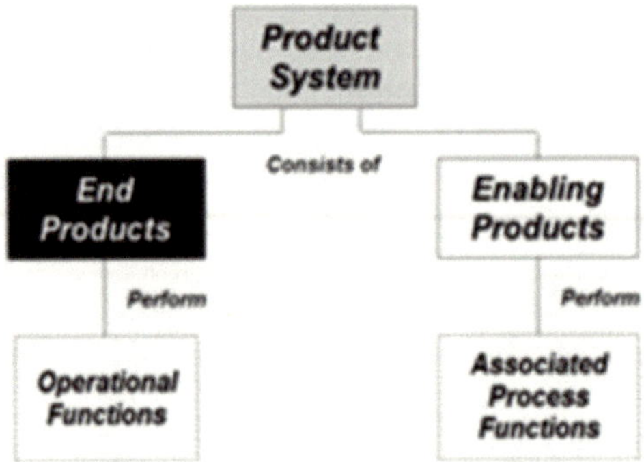

Figure 2.
Product System Components [7].

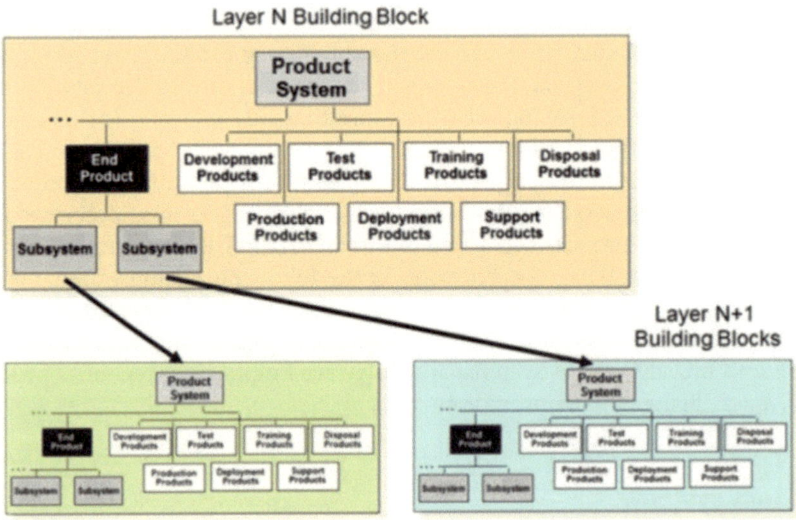

Figure 3.
End products and enabling products [4].

enabling products are complex or their relationship to the end product is complex. Otherwise, the use of conventional product development process is sufficient [5].

The processes of the SE activities are used to develop and realize the end products. Fourteen levels of technical processes are described as SE activities for product realization in ISO/IEC 15288 standards [5]. These processes are Business and Mission Analysis Process, Stakeholder Analysis and Requirements Identification Process, Identification of Systems Requirements Process, Architecture Definition Process, Design Definition Process, System Analysis Process, Implementation Process, Integration Process, Verification Process, Transition Process, Validation Process, Operation Process, Maintenance Process, and Disposal Process.

For product systems, product realization processes are applied to each operational/mission product in the system structure starting from the lowest-level product and working up to higher-level integrated products. These processes are used to create the design solution for each product (e.g., by the Product Implementation or Product Integration Process) and to verify, validate, and transition up to the next hierarchical level products that satisfy their design solutions and meet stakeholder expectations as a function of the applicable life cycle phases [8].

3. Enterprise systems

Enterprise systems are complex, highly integrated systems comprised of processes, organizations, information, and supporting technologies, with multifaceted interdependencies and interrelationships across their boundaries. Understanding the SE and management of complex social, technical, and infrastructure dimensions of an enterprise is critical to achieving and sustaining the enterprise performance. This section attempts to address the following questions:

- What are the key attributes of a successful enterprise, both today and emerging?
- What are the key concepts, elements, and interrelationships that comprise an enterprise system?
- What is involved in "architecting" and "engineering" an enterprise to achieve the desired characteristics in the context of environment, business model, and associated product system? [9]

Enterprises are studied by social scientists for many years; however, according to this study, all of the perspectives against the Enterprises were only concentrated on organizational structure or the information architecture [4].

Enterprise-type projects need to be looked as a system of systems (SOS), rather than as a collection of functions connected solely by information systems and shared facilities. What differentiates the design of an enterprise system from product system lies in the inclusion of people and organization (including policies, leadership, and facility) as part of the system, but not just only including as a user or operator of the s.

Most business enterprises include one or more SoSs. For instance, most businesses have integrated many of their back office systems such as employee systems, payroll systems, and accounting systems. In addition they may also have an integrated set of customer facing systems such as order of entry, pricing systems, billing, service monitoring, inventory management, and customer help. These types of SoS tend to be relatively static in that the systems are always linked and interoperating with each other to support the organization's key business functions [6].

Enterprise acquires or develops systems or individual elements of a system. The enterprise can also create, supply, use, and operate systems, SOS, or system elements. Since there can possibly be several organizations involved in this enterprise venture, each organization can be responsible for particular systems or certain kinds of elements. Each organization brings their own organizational capability with them, and the unique combination of these organizations leads to the overall operational capability of the whole enterprise. These concepts are illustrated in **Figure 4** [4].

The primary purpose of an enterprise is to create value for society, other stakeholders, and for the organizations that participate in that enterprise. This is illustrated in **Figure 3** that shows all the key elements that contribute to this value creation process. These elements in the enterprise can be treated as a system with SOS perspective, where the processes, methods, and tools of SE activities can be applied to.

There are three types of organizations of interest: business, projects, and teams. A typical business participates in multiple enterprises through its portfolio of

Figure 4.
Individual competence leads to organizational, system, and operational capability [4].

projects. Large SE projects can be considered as an enterprise in their own merit, with participation by many different business areas, and may be organized as a number of sub-projects.

Enterprise system cannot be separated from their environment and its associated product system. In studying any engineering systems, it can be seen that the product system (i.e., what gets delivered to the stakeholders) and the enterprise system are truly intertwined, along with the overall environment in which they reside. An established enterprise will innately influence any new product system that is produced by that enterprise. Similarly, the characteristics of a product system to be developed will drive the enterprise architecture toward a particular structure and set of behaviors. Moreover, in enterprise architecting, we are faced with an important consideration: how do you architect an enterprise that can most effectively produce a desired "product system"? Today we can, at best, cite heuristics and emerging principles on how enterprises should be architected. Current research in enterprise systems architecting is working toward transforming enterprise architecting from an art to a science, wherein enterprises can be predictably architected and engineered [4].

4. System of systems

The term "System of Systems" (SOS) has been used since the 1950s to define the systems that are composed of independent subsystems, which behave jointly toward a common goal through the synergism between them. For instance, energy systems, transport, production, government, and military enterprises are examples of SoS [6].

Understanding the environment in which a system or SoS is generated and established is central to understanding how best to apply SE principles within that environment. Differences between individual or subsystem observations and SoS are given in **Table 1** [10].

Today's requirement for more complex, more capable systems in short time is leading many organizations to the integration of new and existing systems with commercial off-the-shelf (COTS) products into network-centric, knowledge-based system of systems. With this method system, development activities to make up the new multisystem architecture, identify sources to either supply or develop the

Aspect of Environment	System	Acknowledged System of Systems
Management & Oversight		
Stakeholder Involvement	Clearer set of stakeholders	Stakeholders at both system level and SoS levels (including the system owners), with competing interests and priorities; in some cases, the system stakeholder has no vested interest in the SoS; all stakeholders may not be recognized
Governance	Aligned PM and funding	Added levels of complexity due to management and funding for both the SoS and individual systems; SoS does not have authority over all the systems
Operational Environment		
Operational Focus	Designed and developed to meet operational objectives	Called upon to meet a set of operational objectives using systems whose objectives may or may not align with the SoS objectives
Implementation		
Acquisition	Aligned to ACAT Milestones, documented requirements, SE with a Systems Engineering Plan (SEP)	Added complexity due to multiple system lifecycles across acquisition programs, involving legacy systems, systems under development, new developments, and technology insertion; Typically have stated capability objectives upfront which may need to be translated into formal requirements
Test & Evaluation	Test and evaluation of the system is generally possible	Testing is more challenging due to the difficulty of synchronizing across multiple systems' life cycles; given the complexity of all the moving parts and potential for unintended consequences
Engineering & Design Considerations		
Boundaries and Interfaces	Focuses on boundaries and interfaces for the single system	Focus on identifying the systems that contribute to the SoS objectives and enabling the flow of data, control and functionality across the SoS while balancing needs of the systems
Performance & Behavior	Performance of the system to meet specified objectives	Performance across the SoS that satisfies SoS user capability needs while balancing needs of the systems

Table 1.
Comparing system and acknowledged system of systems [10].

necessary components, and also integrate and test the high-level components are in development and are being defined as SoS Engineering. Many people believe that software-intensive systems are widely found in human use. It is known that the software is embedded in automobiles, household appliances, and even computers and sensors on bicycles to tell us how far we have gone and our average speed. It is also easy to see, once one understands the concepts of SoS that SoS can also be found anywhere [11].

5. SE technical processes for enterprise systems

Enterprise SE is defined as the body of knowledge, principles, and practices having to do with the analysis, design, implementation, and operation of an enterprise. In a continued changing and competitive environment, systems engineers who work on enterprise projects should consider a fundamental question: "How to design and improve all system elements associated with the enterprise through the use of SE and analysis methods and tools more effectively for achieving its goals and objectives." In Enterprise SE, there exist three schools of thoughts:

- The enterprise can be viewed as a complex system.
- The enterprise is to be viewed as a system of processes that can be engineered both individually and holistically.
- The use of engineering rigor in transforming the enterprise.

In the enterprise engineering paradigm, the enterprise is viewed as a complex system of processes that can be engineered to accomplish specific organizational objectives. Enterprise systems engineering recognizes the ever-changing organic nature of the enterprise and therefore has a valid worldview or paradigm.

Attempts to define Enterprise SE frequently fall back on refining previous concepts of systems integration and interoperability rather than on Enterprise SE as a whole; indeed, refining all of these concepts is useful, yet the focus is still on modeling and integrating already-existing systems or components. Because it is already stated in literature of Enterprise SE that effective enterprise integration involves not only hardware, equipment, and data but also people, technology, and business processes [12].

Enterprise SE systems thinking should cover the following aspects:

- Conceptual Foundation for Enterprise SE: Considering and classifying types of enterprises and their characteristics, both as in hardware and software; drawing upon the related fields of complexity science and sociology to explain and predict the interactions between processes and the behavior of enterprises undergoing in change

- Enterprise Technical Strategies: Enterprise architecture and associated frameworks (TOGAF, DODAF, Zachmann, etc.); service-oriented architectures, framework architectures, and components of systems; generic platform architectures crossing multiple systems; reusable architectural patterns; strategic management of technology; and integrated technical support environments, covering modeling, integration, and organization-level information and sharing

- Enterprise Process and Management: Combination of enterprise management with organizational objectives; enterprise systems of systems engineering; agile development methods to counter uncertainty; management of programs and portfolios integrated with appropriate engineering processes; commercial policies and contract management and decision-making techniques at all levels to support the above

- Organizational Design and Change Management: Shaping the organization for adaptability and to match the types of complexity involved; top down vs. peer-to-peer collaboration and governance strategies; enterprise leadership styles; implementing; sustaining and measuring change

Enterprise systems architecture is a new strategic view that creates a systems perspective, viewing the entire enterprise as a complex system encompassing multiple views such as organization view, process view, and knowledge view and enabling information technology view in an integrated framework.

According to the MITRE [9], the following process areas are applied to close the gap between the Enterprise SE and Product SE:

- Strategic technical planning

- Enterprise architecture modeling

- Capability-based planning analysis

- Technology planning

- Enterprise analysis and assessment

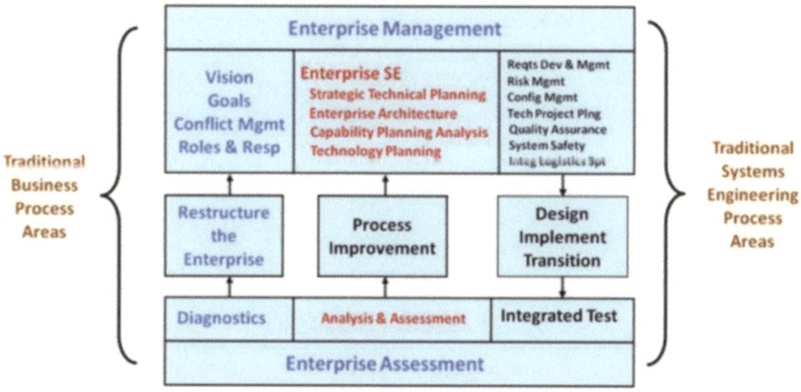

Figure 5.
Enterprise systems engineering process areas in the context of the entire enterprise [13].

The Enterprise SE process is shown in the context of the entire enterprise in **Figure 4** [13]. The Enterprise SE processes are depicted in the middle with the business processes on the left and technical systems engineering processes on the right. In **Figure 5**, these business processes are described as in relation to the business activities.

For HAVELSAN study projects discussed in this chapter, the following table is constructed according to the needs of the enterprise projects, including technical and SE processes, and product development project systems engineering and technical processes that are not meeting the Enterprise scope of work. **Table 2** provides the basis for comparison of input/output and processes of the product, service, and enterprise SE technical processes according to the INCOSE guidelines and ISO/IEC 15288 standards.

Process no.	Project type technical processes	Product development project systems engineering technical process and outputs	Enterprise system project type systems engineering technical process and outputs
1	Business and mission analysis process	• Statement of work • Technical specifications • Technical solution documents • Engineering analysis (trade-off analysis) • Regulations and/or business processes • National and international standards • Concept of operations document • Customer meeting records	• Statement of work • Technical specifications • Technical solution documents • Engineering analysis (trade-off analysis) • Regulations and/or business processes • National and international technical standards • Customer meeting records • Organizational strategic plan • Capability-based planning analysis • Technology standards planning
2	Stakeholder analysis and requirement identification process	• Project plans • Work breakdown structures • Project feasibility estimations	• Program management plan • Work breakdown structure • Project feasibility • Opportunity-risk evaluation plan • Enterprise architecture roadmap process
3	Identification of systems requirement process	• System/subsystem specifications • System/subsystem interface specifications • Software specifications	• Enterprise system/subsystem specifications • Enterprise system/subsystem interface specifications

Process no.	Project type technical processes	Product development project systems engineering technical process and outputs	Enterprise system project type systems engineering technical process and outputs
		• Hardware specifications • Hardware interface specifications • System/subsystem requirements review • Software requirements review • Hardware requirements review • Test plan • Validation and verification plan	• System/subsystem requirements review • Validation and verification plan
4	Architecture definition process	• System/subsystem design model (SysML/UML, system IT models, etc.) • Data configuration lists • Interface design model (SysML/UML, etc.) • Software design model (UML, etc.) • Data model (UML) • Database design model (UML, etc.) • Hardware design model (AutoCad mechanical and design drawings, Ansys or CATIA 3D models and engineering analysis, cable/wire drawings, mass model, structure model, electromagnetic model analysis, RF propagation, communication link budget analysis, etc.) • Engineering analysis reports (trade-off analysis, safety, ergonomy, security, reliability, maintainability, survivability, availability, material and process analysis)	• Enterprise architecture design model (based on TOGAF or DODAF standards, etc.) • Enterprise architecture interface design model (based on TOGAF or DODAF standards) • Enterprise architecture configuration data list • Enterprise architecture data model (SysML/UML, etc.) • Organizational enterprise architecture system model (SysML/UML, etc.) • Enterprise architecture operation model (BPM, etc.) • Enterprise architecture organization model Kurumsal (SysML/UML, etc.) • Engineering analysis reports (enterprise architecture trade-off analysis, enterprise architecture service level performance calculations, enterprise architecture service level time and human source analysis, enterprise architecture operation security analysis, service flow simulation and calculations)
5	Design definition process	• System/subsystem design review • Interface design description • Software design description • Database design description • Hardware design description • Primary item development specification	• Enterprise architecture system/subsystem design document • Enterprise architecture system interface document • Enterprise architecture program management plan • Architecture product project portfolio project management plan • Enterprise architecture service system project portfolio project management plan • Enterprise architecture system engineering management plan • Enterprise architecture roadmapping • Acquisition plan • Technical specification • Administrative specification

Process no.	Project type technical processes	Product development project systems engineering technical process and outputs	Enterprise system project type systems engineering technical process and outputs
			• Statement of work • Configuration management plan
6	System analysis process	• System/subsystem design review (SSDR) • Software preliminary design review (SW-PDR) • Software critical design review (SW-CDR) • Hardware preliminary design review (HW-PDR) • Hardware critical design review (HW-CDR)	• Enterprise architecture system/ subsystem design review • Enterprise architecture operational design review • Enterprise architecture roadmap design review • Strategic plan analysis • Capability-based technical planning analysis • Technology and standard planning analysis • Enterprise architecture technical product development project portfolio design reviews • Enterprise architecture service system project portfolio design reviews • Subcontractor or project plan peer reviews • Statement of work peer reviews • Administrative specification peer reviews
7	Implementation process	• Systems engineering management plan • Software development plan • Hardware development plan • Coding standards • Prime item product specification • Technical design drawings and part lists • Development process adaptation document • Project environment evaluation report	• Enterprise architecture program management plan • Enterprise architecture technical system product development portfolio project management plan and systems engineering process • Enterprise architecture program service system project portfolio project management plan and service systems engineering process • R&D project portfolio for enterprise architecture systems • Systems engineering plan for enterprise systems • Strategic technical plan implementation analysis checklist • Capability-based implementation analysis checklists • Technology and standard planning checklists • Enterprise architecture measurement and evaluation metrics and checklists
8	Integration process	• Validation and verification plan • Test plan • Test description document	• Enterprise architecture program management plan

Process no.	Project type technical processes	Product development project systems engineering technical process and outputs	Enterprise system project type systems engineering technical process and outputs
		• Test procedure • Test report • Peer review planning and tracking list	• Enterprise architecture systems engineering management plan • Verification and validation control plan • Enterprise architecture system test description document • Enterprise architecture system test procedure • Enterprise architecture system test report • Peer review planning and tracking list • Product development integration validation process for technical systems in enterprise architecture • Strategic technical implementation analysis checklist • Capability-based implementation analysis checklists • Checklist leri technology and standard plans • Enterprise architecture measurement and evaluation metrics and checklists
9	Verification process	• Verification and validation control plan • Test readiness review • Test plan • Test description document • Test procedure • Test report • Version description document • Functional and physical configuration audit	• Enterprise architecture program management plan • Systems engineering management plan • Verification and validation control plan • Test description document • Test procedure • Test report • Version description document • Functional and physical configuration audit • Enterprise architecture technical product development or acquisition project portfolio integration verification process • Strategic technical verification of realization analysis checklists • Capability-based implementation analysis checklists • Technology and standards planning checklists • Enterprise architecture measurement and assessment checklists
10	Transition process	• Software installation plan • Hardware installation plan • Software transition plan • Installation requirements	• Enterprise architecture roadmap

Process no.	Project type technical processes	Product development project systems engineering technical process and outputs	Enterprise system project type systems engineering technical process and outputs
		• Transition to installation and operation plan • Transition to installation and operation report • Migration plan (for construction and infrastructure operations) • Occupational health and safety plan (for construction and infrastructure operations) • Product specification • Software product specification • User manual • User training plan	• Enterprise architecture technical products systems engineering transition process • Enterprise architecture technical Kurumsal Mimari service systems engineering transition process • Implementation of strategical technical analysis checklists • Implementation of capability-based analysis checklists • Implementation of technology and standards planning checklists • Enterprise architecture measurement and analysis checklists
11	Validation process	• Verification and validation plan • Test readiness review • Test plan • Test description document • Test procedure • Test report • Version description document • Functional and physical configuration audit	• Enterprise architecture roadmap • Program management plan • Data configuration list • Systems engineering management plan • Validation and verification plan • Test readiness review • Test plan • Test description document • Test procedure • Test report • Operation technical plan checklist • Version description document • Functional and physical configuration audit • Implementation of strategical technical analysis checklists • Implementation of capability-based analysis checklists • Implementation of technology and standards planning checklists • Enterprise architecture measurement and analysis checklists
12	Operation process	• Installation requirements • Transition to installation and operation plan • Transition to installation and operation report • Work standards • Application procedure	• Program management plan • Systems engineering management plan • Enterprise architecture roadmapping • Enterprise architecture technical system product systems engineering operation process • Enterprise architecture service systems engineering operation process

Process no.	Project type technical processes	Product development project systems engineering technical process and outputs	Enterprise system project type systems engineering technical process and outputs
			• Capability-based implementation analysis checklists • Technology and standards planning checklists • Enterprise architecture measurement and evaluation metrics checklists
13	Maintenance process	• System/subsystem requirements specification • System/subsystem design document • Software maintenance plan • Hardware maintenance plan • Work standard-isolation of faults/removal	• Program roadmapping • Systems engineering management plan • Enterprise architecture roadmapping • Enterprise architecture technical systems product development maintenance process • Technology and standards implementation checklists • Enterprise architecture measurement and evaluation checklists
14	Disposal process	• System/subsystem requirements review • System/subsystem design document • Project management plan or environment management plan • Systems engineering management plan • Atık Takip Formu • Environmental effect analysis • Environmental management program • Disposal water tracking form	• Program management plan • Systems engineering management plan • Configuration management plan • Enterprise roadmapping • Strategic technical analysis • Capability-based implementation analysis • Technology and standards implementation report • Enterprise measurement and evaluation report

Table 2.
ISO/IEC 15288 system standards and technical process for enterprise systems.

6. Conclusion

In conclusion, several remarks can be made based on the results presented in **Table 2**. **Table 2** defines the differences in the system types and explains technical engineering life cycles for each system type. The table provides data to compare the technical engineering life cycle process's inputs and outputs according to the INCOSE SE book and ISO/IEC 15288 standards. Enterprise systems of systems engineering applies systems engineering fundamentals to the design of an enterprise. It is created by knowledge, principles, and processes tailored to the design of enterprise systems. Enterprise is a complex, socio-technical system that includes interdependent resources of people, information, and technology to reach a common goal. Enterprise systems engineering is needed when the complexity is faced which breaks down the assumptions upon which textbook systems engineering is based, such requirements being stable and well understood; a system configuration can be controlled among the different stakeholders. There are four different steps in Enterprise systems process; it includes technology planning (TP), capabilities-based engineering analysis (CBEA), enterprise architecture, and enterprise analysis and assessment.

At HAVELSAN Inc., for software product development projects, the CMMI V3-based Development product life cycle management is used. For systems development process, the ISO/IEC 15288 Systems Engineering standards are used. However, because of the structural differences required by the project's needs for a specific system type, it is necessary to customize and develop new SE and technical approaches for the required products and services of enterprise system projects. For instance, in our Information and Security Technologies Division at HAVELSAN Inc., there is 30% or more projects classified as Service Systems, about 50% classified as product development projects, about 4% as enterprise systems projects, and 16% is classified as combination of the three mentioned systems. Our current product development technical processes are not useful enough to use them in enterprise system projects. HAVELSAN has devised a new methodology of technical systems engineering processes for the service and enterprise systems projects to accomplish customer needs and required quality assurance. Enterprise system projects require enterprise resources such as people, processes, infrastructure, and strategic objectives. To fulfill the enterprise system project requirements related to enterprise resources, one needs to define new SE and technical engineering processes. **Table 1** will help system engineers to compare the enterprise system project perspectives in development stages with the product system development stages.

Author details

Alper Pahsa
HAVELSAN Inc., Ankara, Turkey

*Address all correspondence to: apahsa@havelsan.com.tr

References

[1] David ES, Michael B, Obaid Y. Systems Engineering and Program Management-Trends and Costs for Aircraft and Guided Weapons Programs. United States Air Force; 2006. ISBN: 0-8330-3872-9

[2] Keating C, Rogers R, Unal R, Dryer D, Poza SA, Safford R, et al. Systems of systems engineering. Journal Engineering Management Journal. 2003; 15(3):36-45

[3] Carlock P, Lane AJ. System of Systems Enterprise Systems Engineering, the Enterprise Architecture Management Framework and System of Systems Cost Estimation. SoS, USCCSE; 2006. Article 618

[4] INCOSE, Guide to the Systems Engineering Body of Knowledge (SEBoK), The Trustees of the Stevens Institute of Technology, Stevens, Part-4, INCOSE BKCASE, v1.7 2016

[5] Cardoso J, Schmitz B, Kieninger A. Service Design - Service Research. New York, US: Springer International Publishing; 2015. pp. 325-357

[6] Lane JA. What Is System of Systems and Why Should I Care? Daniel J. Epstein Department of Industrial and systems Engineering, University of Southern California, USC, CSSE-2013; 2013

[7] ANSI/EIA. EIA-632, Processes for Engineering a System. Philadelphia, PA, USA: American National Standards Institute (ANSI)/Electronic Industries Association (EIA); Government Electronics and Information Technology Association Engineering Department. VA, US; 2003

[8] NASA. Systems Engineering Handbook, NASA/SP-2007-6105 Rev1. Washington DC, US: NASA Headquarters; 2007

[9] Rebovich G Jr. Volume 2: Systems Thinking for the Enterprise: New and Emerging Perspectives. MITRE Corporation; 2005

[10] Nielsen BC, Larsen GP, Fitzgerald J, Woodcock J, Peleska J. Systems of systems engineering: basic concepts, model-based techniques, and research directions. ACM Computing Surveys. 2015;**48**(2):18

[11] Systems and Software Engineering Director, Systems Engineering Guide for Systems of Systems, Deputy Under Secretary of Defense (Acquisition and Technology Office of the Under Secretary of Defense (Acquisition, Technology and Logistics)), Version 1; 2008

[12] Saenz AO, Ronald EG, Centeno M. Defining enterprise systems engineering. International Journal of Industrial and Systems Engineering. 2009. DOI: 10.1504/IJISE.2009.024155

[13] DeRosa JK. Enterprise Systems Engineering. Presented at Air Force Association, Industry Day, Day 1, August 4, 2005; Danvers, MA, USA; 2005

Section 3

SOS Architecture Design and Analysis

Chapter 4

SOS Enterprise, SOSE CONOPS, SOSE Architecture Design Approach: A Perspective on Space and Airborne Systems

Tien M. Nguyen

Abstract

The objective of this chapter is to (i) define System-of-Systems Enterprise (SOSE), SOSE Concept of Operations (CONOPS), and SOSE Architecture (SOSEA) CONOPS assessment, and (ii) discuss their differences using examples from existing space and airborne systems. The chapter also describes the SOS design challenges and presents an SOSE Architecture design approach addressing these challenges. In addition, DOD Architecture Framework Version 2.02 (DODAF-v2.02) views will be discussed along with a recommendation for a set of key DODAF views to capture system architecture artifacts with practical examples involving SOS Enterprise architectures for notional space-based communications system and manned airborne Intelligence, Surveillance, and Reconnaissance (ISR) platform.

Keywords: SOS Enterprise (SOSE), SOSE CONOPS, SOSE Architecture (SOSEA) design, SOSE integration, compatibility matrix, SOSEA evaluation metrics, SOSE capability component, capability gap analysis framework, capability management framework, requirements-based, capability-based, integration analysis framework

1. Introduction

Currently, from a combined space and airborne perspective, one can categorize existing "enterprises" as: (i) military enterprise, (ii) civilian enterprise, and (iii) commercial enterprise for space and airborne applications. This chapter uses existing U.S. Department of Defense (DOD) and The International Council on Systems Engineering (INCOSE) definitions on System-of-Systems (SOS) and Family-of-Systems (FOS) [1–3] to define these enterprises through the use of practical examples and design scenarios. **Figure 1** illustrates the space SOSE concept in general. As an example, current commercial space enterprise consists of (i) FOS of Broadcasting Satellites (FOS-BS), (ii) FOS of Wideband Internet Satellites (FOS-WIS), and (iii) FOS of Data, Video, Audio Communications Satellites (FOS-DVACS). For commercial space enterprise, the SOS environment can be defined as:

- A set of connections among satellites within each FOS, namely, FOS-BS, FOS-WIS, and FOS-DVACS
- A set of connections among FOS-BS, FOS-WIS, and FOS-DVACS.

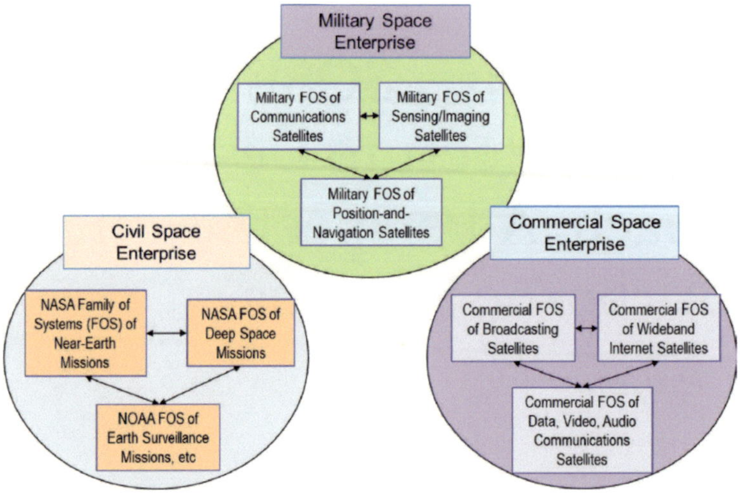

Figure 1.
Space SOS Enterprise definition.

Similarly, the definition for military space enterprise includes: (i) FOS of communications satellites, (ii) FOS of sensing/imaging satellites, and (iii) FOS of Position-Navigation-and-Timing (PNT) satellites. For civilian space enterprise, it includes: (i) FOS for near-Earth missions, (ii) FOS for deep space missions, and (iii) FOS for earth surveillance missions. The same definitions can be derived for the airborne enterprises.

As the space and airborne operational environment is moving from "system" to "SOS" environment, the system (or SOS) engineers face many challenges when designing a new system that can operate within a SOS environment. Some of the key challenges are:

- Backward compatibility: Compatible with existing SOS environment (also referred to as "As-Is");

- Forward compatibility: Compatible with future planned SOS environment (or "To-Be");

- System architecture design is very challenging in SOS environment because of the transition from "requirement-based" to "capability-based" to avoid stovepipe solution. For a stovepipe requirement-based system design, a set of technology enablers[1] is chosen at the time of the design, and the system requirements are derived from that selected set of technology enablers and then they are flown-down to subsystems and associated hardware, software, and middleware components. For requirement-based system design, the architecture trade space is well defined due to known technology enablers. But for capability-based system design, the architecture trade space is unknown to systems designers due to unknown technology enablers. The following challenges described below are the by-products of this transition;

[1] Technology Enabler (TE) is defined as the technology that concretely fulfills the required user's capabilities (or warfighter's capabilities for military systems). Examples of TE for space applications are: Software Defined Radio (SDR) modem, Beamforming network, Digital Channelizer and Beamformer (DCB), Adaptive Linearizer, etc.

- Technology enablers identification to ensure synchronization with the required system design features in SOS environment.
- Capability gap analysis: Analyzing capability gaps and generating solution to fill the gap should be synchronized with changing customer's needs.
- SOS integration analysis to ensure early identification of potential integration issues between the new system with existing SOS environment.
- SOS capabilities management ensuring synchronization between existing SOS's requirements with planned capabilities.

- A robust, agile, and flexible SOSE Architecture design approach leading to a system architecture solution that (i) meets customer's requirements for seamless interfacing with both existing/legacy and future systems, and (ii) can identify desired architecture products early during the concept design cycle allowing for accurate costing of the system.

This chapter will describe a SOSEA design approach that can address the above challenges. The chapter is organized as follows: (i) Section 2 describes SOSE CONOPS, SOSEA alternative solutions, and associated SOSEA evaluation metrics; (ii) Section 3 discusses U.S.DODAF, SOSEA perspectives across enterprise domains, SOSE capability components, and SOSE integrated capability. This section also presents a perspective on enterprise domains, including Doctrine (D), Organization (O), Training (T), material (m), Leadership (L), Personnel (P), Facility (F), and Policy (P), abbreviated as DOTmLPF-P; (iii) Section 4 proposes an approach for the design of a system in a SOSE environment that can address all the challenges discussed above; (iv) Section 5 provides examples on notional SOSEA alternative solutions using the proposed approach described in Section 4 for typical space-based communications system and manned airborne Intelligence Surveillance Reconnaissance (ISR) platform; and (v) Section 6 concludes the chapter with remarks on future design and analysis of SOSE space and airborne systems.

2. SOSE CONOPS and SOSEA alternative solutions

Section 2.1 describes SOSE CONOPS, Section 2.2 discusses SOSEA alternative solutions, and Section 2.3 proposes a potential set of SOSEA assessment metrics for evaluating alternative architecture solutions.

2.1 SOSE CONOPS description

This subsection provides a description of SOSE CONOPS through an example using the space SOSE definition presented in Section 1. **Figure 2** presents a notional SOSE CONOPS for Satellite Communication (SATCOM) applications. The figure illustrates the connections among military, civilian, and commercial SOS Enterprises. This SOSE CONOPS shows: (i) Military satellite Node 1 (MIL-Node 1), MIL-Node 2, and MIL-Node 3 communicating with Military Gateway 1 (MIL GW 1), MIL GW 2, MIL GW 3, User Mobile Terminal 1, and User Mobile Terminal 2; (ii) MIL-Node 2 communicating with MIL GW 1 and MIL GW 2 through the datalinks 1 and 2, respectively; (iii) Commercial Satellite Node 4 talking to Commercial Satellite Gateway 1 through RF signal 2; and (iii) Civilian Satellite Node 5 talking to Civilian Satellite Gateway 2 through RF signal 1. For this notional SOSE CONOPS, the RF

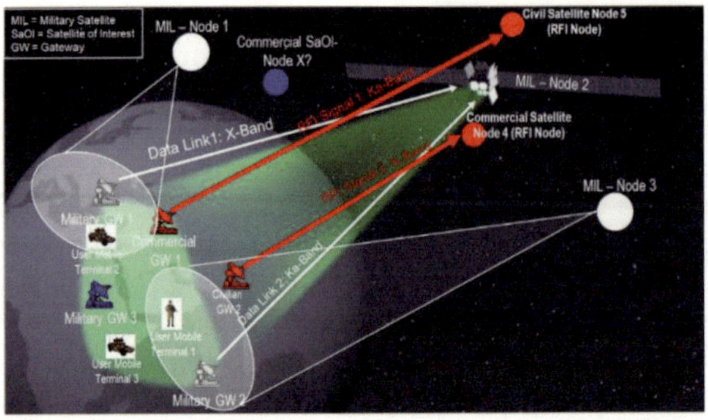

Figure 2.
Notional SOSE CONOPS for SATCOM applications.

signals 1 and 2 are interfering with the datalinks 1 and 2, hence the Commercial Satellite Node 4 and Civilian Satellite Node 5 are considered as Radio Frequency Interference (RFI) nodes. As shown in this SOSE CONOPS, the SOSEA design issue here is to identify an existing Commercial Satellite Node, represented by a commercial Satellite of Interest (SaOI) Node X, that can enhance the resiliency[2] of the military network consisting of MIL-Node 1, MIL-Node 2, and MIL-Node 3 [4].

2.2 SOSEA alternative solutions description

Using similar approach presented in Section 2.1, this section defines the concept of SOSEA alternative solutions using space SOSE examples. For the SOSE CONOPS presented in **Figure 2**, there are many possible SOSEA alternative solutions based on existing commercial satellites' availability. As an example, a quick survey was conducted to collect data for this notional construction of a SOSE commercial space database, the following commercial satellites and Ground Stations (GS) are found to be available:

- SPACEWAY Satellites 2 and 3 and ARSAT satellite 2; there is a total of three commercial satellites available, namely, Commercial Satellite Nodes 6, 7, and 8, respectively.
- U.S. Universal Space Network (USN) GS in California and U.S. Orbital Tracking Corp GS in California; there is a total of two GS available, namely, commercial GS Nodes 3 and 4, respectively.

For the notional SOSE CONOPS described in **Figure 2**, there are six possible (3 × 2) SOSEA alternative solutions. **Figure 3** describes a process for deriving a set of potential SOSEA alternative solutions for this example. Thus, a system architecture solution optimization is dependent on:

- SOSE environment, that is, depending on the SOSE CONOPS.
- SOSEA alternative solutions. As explained above, for this notional CONOPS, there are six potential SOSEA alternative solutions.

---

[2] U.S. DOD uses Resilient Capacity addressing Avoidance, Robustness, Reconstitution, and Recovery [4]. Recently, the author has introduced two additional resiliency metrics addressing RFI problems. The first metric is Resilience Assessment Index against RFI (RAI-RFI) and the second metric is Spectrum Resiliency Assessment Index (SRAI) [8, 9].

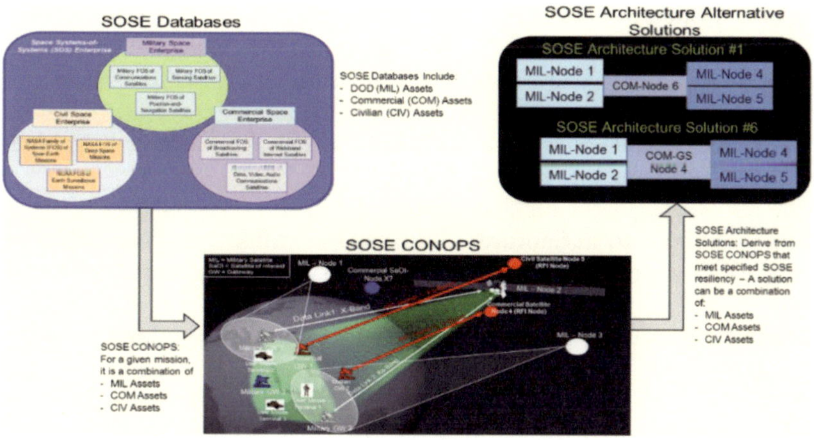

Figure 3.
A process for generating SOSE Architecture alternatives.

Architecture TPMs	Desired Criteria	Meeting Criteria? (Y/N)
Critical Operational Issue (COI) # 1: Does the SOSE solution improve the warfighters' ability to produce *timely* and *accurate* target coordinates to support customer missions? *(Operational Viability Issue)*	Notional SoSE immediate needs to support Customer planning: - E.g., meeting customer Initial Capability Document (ICD) (when the systems are completely transformed to capability-based, this document becomes obsolete) - Work with customer to define the needs Notional Criteria for meeting timely and accurate target coordinates (COI # 1) are: - E.g., Meeting customer ICD	
COI # 2: Is the proposed SOS solution suitable for use with existing SOSE services and joint C4ISR assets? *(Technology Feasibility Issue)*	Notional Criteria for meeting COI # 2 are: - Compatibility matrix: interfaces and standards are compatible among required HW/SW, SOSE and joint C4ISR assets (depending on the selected CONOPS) - Connectivity matrix: Connectivity among all nodes (depending on the selected CONOPS)	
COI # 3: Does the required SOSE solution provide capability to improve integration between SOSE and system level? *(Operational Viability Issue)*	Notional Criteria for meeting COI # 3 are: - Seamless flow from SOSE to system and vice versa - Do we have enough communications bandwidth? - Does it meet current and future system/subsystem interface standards requirements?	
Critical Integration/Implementation Issues (CIII): Does the required SOSE solution meet the schedule and cost constraints? *(Affordability Issue)*	Notional Criteria for meeting CIII are: - Schedule: Required HW/SW schedule vs. Master schedule - Cost: Minimum implementation cost and/or reduced Total Ownership Cost (TOC) and meeting the allocated budget	
Critical SOSE MOEs: Does the required SOSE solution improve KPPs, Resiliency, Kill Rate (KR), Life-Time Growth of the Target (LTGT), and SOS Efficiency? *(Operational Viability Issue)*	Notional Criteria for meeting SOSE Measure of Effectiveness (MOEs) Specified in the ICD are: - E.g., enhance Resilient Capacity by 25% - E.g., the current average KR is about 0.1% - The current average LTGT is TBD - The current SOS Efficiency, Ae, is TBD - The current set of KPPs (Key Performance Parameters) is TBD	

Table 1.
Notional SOSEA technical performance metrics (TPMs) for military space applications.

2.3 Proposed SOSEA evaluation metrics

Defining SOSE Architecture evaluation metrics, such as Quality-of-Service (QoS) or Measure of Goodness (MoG), for assessing alternative system architecture solutions in a complex SOS environment is not a trivial task. The QOS of a system architecture solution can be defined in terms of the Resilient Capacity (see Footnote 3) for a complex military space systems network or a Package Error Rate (PER) for a commercial satellite systems network. Based on past experience, **Table 1** presents a notional set of metrics for typical space SOS military applications.[3] This set of metrics can be used for assessing MoG of the six notional SOSE military architecture alternative solutions discussed in Section 2.2. The set of metrics presented in **Table 1** emphasizes

[3] The tailoring of this metrics for civilian and military applications is straight forward.

on the Critical Operational Issues (COIs) for meeting the users' (or warfighter's) needs. One of the key COIs is the "Technology Feasibility Issue" that addresses the compatibility issues of the new system with existing/legacy SOSE services and joint Command, Control, Communications, Computers, Intelligence, Surveillance, and Reconnaissance (C4ISR) assets, see COI #2 of **Table 1**. As indicated in **Table 1**, the notional criteria for meeting the COI # 2 are:

- Compatibility matrix: Captures interfaces and industry standards that are compatible among required HW/SW, SOSE systems, and joint C4ISR assets;
- Connectivity matrix: Captures connectivities among all nodes.

3. SOSEA perspectives and capability view

This section is organized as follows: Section 3.1 describes the current U.S. DODAF Version 2.02 artifacts; Section 3.2 presents a SOSEA perspective for commercial and civilian applications; Section 3.2 presents a SOSEA perspective for military applications; and Section 3.3 discusses SOSE capability components and SOSE integrated capability.

3.1 U.S. DODAF artifacts

DoDAF Version 2.02 is the most current version for the U.S. department of defense [3]. For all U.S. DOD programs, the architecture components are expected to conform to DoDAF artifacts to the maximum extent possible. The conformance ensures the reuse of information, architecture artifacts, models, and viewpoints can be shared with common understanding [3]. DODAF artifacts include EIGHT viewpoints [3]:

i. All Viewpoint (AV): Describes the overarching aspects of architecture context that relate to all viewpoints.

ii. Capability Viewpoint (CV): Describes capability requirements, the delivery timing, and the deployed capability.

iii. Data and Information Viewpoint (DIV): Describes data relationships and alignment structures in the architecture content for the capability and operational requirements, system engineering processes, and systems and services.

iv. Operational Viewpoint (OV): Describes operational scenarios, activities, and requirements that support capabilities.

v. Project Viewpoint (PV): Describes the relationships between operational and capability requirements and the various projects being implemented. PV also details dependencies among capability and operational requirements, system engineering processes, systems design, and services design within the Defense Acquisition System process.

vi. Services Viewpoint (SeV): Describes the design for solutions articulating the Performers, Activities, Services, and their Exchanges, providing for or supporting operational and capability functions.

vii. Standards Viewpoint (StdD): Describes the applicable operational, business, technical, and industry policies, standards, guidance, constraints, and forecasts that apply to capability and operational requirements, system engineering processes, and systems.

viii. Systems Viewpoint (SV): For Legacy support, SV describes the design for solutions articulating the systems, their composition, interconnectivity, and context providing for or supporting operational and capability functions.

As described in [3], there can be many DODAF artifacts associated with each view. This chapter proposes a set of key DODAF views and associated artifacts that can be used to provide a systematic way of describing a system architecture in an SOS environment. **Figure 4** proposes an approach to capture system architecture artifacts using DODAF-V2.02 views in a SOS environment.

For examples, as shown in **Figure 4**, OV-1 and OV-4 are used to capture a system or a SOS CONOPS; OV-2, SV-1, and SV-2 for capturing system or a SOS design structure; OV-5, SV-4, SV-4a, SV-4b, SV-5a, SV-5b, and SV-5c for capturing system or SOS operations; and OV-3 and SV-6 for capturing system or a SOS Information Exchange Requirements (IER).

3.2 Civil and commercial perspective: POTmLPF

From a combined civilian and commercial perspective, the enterprise domains usually consist of seven key components, namely, Policy (P), Organization (O), Training (T), material (m), Leadership (L), Personnel (P), and Facility (F). This is also referred to as POTmLPF. A civilian/commercial enterprise solution should address the impacts across these components. The material enterprise solutions address the "m" and "F" components, and the non-material solutions address the "P," "O," "T," "L," and "P" components. **Figure 5** describes a perspective for a commercial enterprise. For civilian enterprise, one replaces "Company" in **Figure 5** by "Civilian Agency" (e.g., NASA or NOAA). This figure shows that a commercial (or civilian) enterprise can be organized across POTmLPF components and these components can be represented by DODAF views as illustrated in **Figure 4**. Furthermore, a commercial enterprise can also be organized by grouping

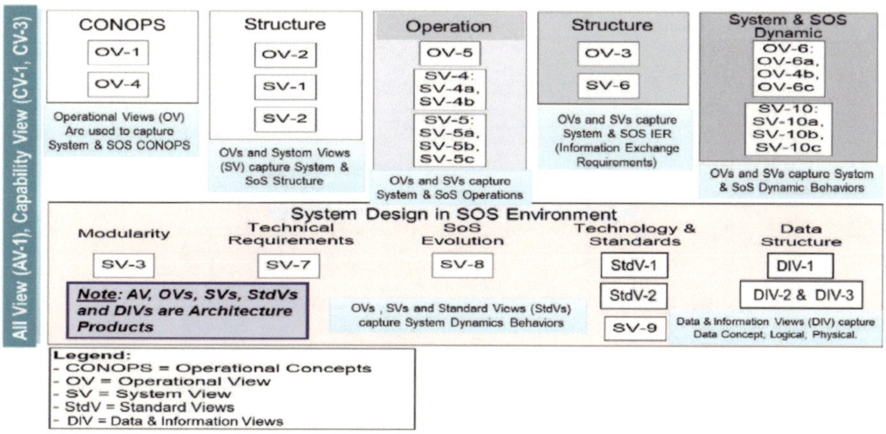

Figure 4.
Proposed DODAF-V2.02 views capturing SOS Architecture.

POTmLPF components into "Company Capability Component #1" through "#N." For example, "Company Capability Component #1" includes POTL components.

3.3 Military perspective: DOTmLPF-P

Military enterprise domain usually consists of eight key components, namely, Doctrine (D), Organization (O), Training (T), material (m), Leadership (L), Personnel (P), Facility (F), and Policy (P). This is also referred to as DOTmLPF-P. **Figure 6** presents a perspective for military enterprise. Similar to commercial enterprise, a military enterprise can also be organized across DOTmLPF-P components, or by grouping DOTmLPF-P components into "Enterprise Capability Component #1" through "#N," and these components can be represented by DODAF views as shown in **Figure 4**.

3.4 SOSE capability view and integrated capability component

As shown in **Figure 6**, for a military SOSE, the SOSE "Capability Component" is defined as a group of any of the DOTmLPF-P components belonging to that

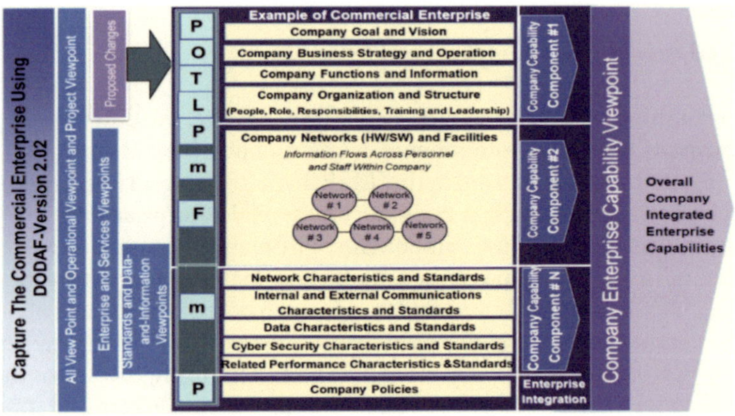

Figure 5.
A commercial SOSE perspective.

Figure 6.
Military SOSE perspective.

enterprise, and that these capability components will contribute to the overall SoSE "Integrated Capabilities Component." The DODAF SOSE Capability View is defined as a view that one can use to graphically describe these SOSE capability components. The DODAF CVs are recommended for capturing these capability components and the overall SOSE integrated capabilities (labeled as "Overall Military Integrated Enterprise Capabilities," **Figure 6**). As an example, the military SOSE capabilities are made up by grouping various capability components across all DOTLmPF-P domains. The architecture models used in the design of a system in a SOS environment shall be developed such that they can (i) generate integrated SoSE capabilities, and (ii) be used to analyze capability gaps. During the design process, these models must be able to generate SOSE capability components that are synchronized with the customer's changing needs. This synchronization process may require consensus among customer's stakeholders to ensure meeting customer's needs through a system life cycle.

As an example, **Figure 7** shows an example of the eight key "SOSE Capability Components" for a notional Airborne Operation Center (AOC). These SOSE Capability Components will be key drivers in the determination of the integrated SOS Capabilities and SOS Requirements for the notional operation center. In general, the SOS Capabilities can be classified into two categories, namely, (i) Existing SOS capabilities (As-Is), and (ii) Future SOS Capabilities (To-Be). Furthermore, the SOS requirements can be classified into material and non-material

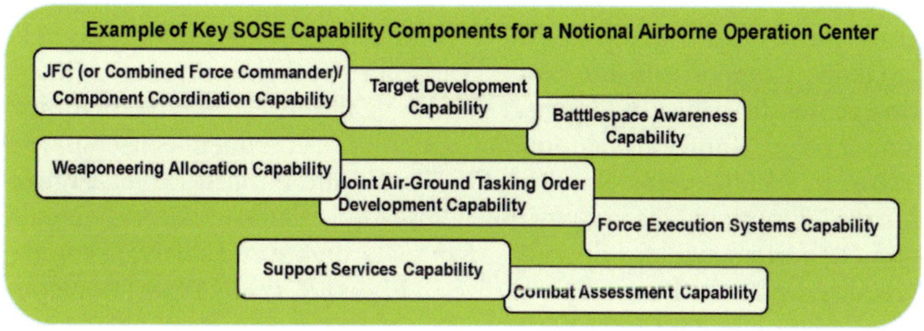

Figure 7.
A capability view for notional military AOC.

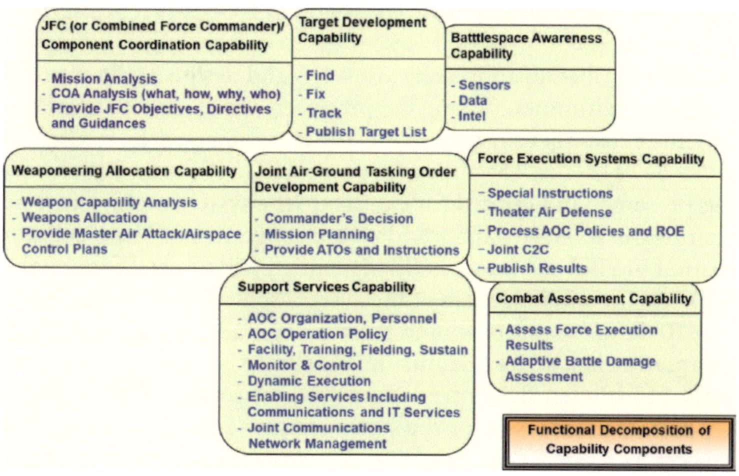

Figure 8.
Example of the functional decomposition of key SOSE capability components to desired capabilities for a notional military AOC.

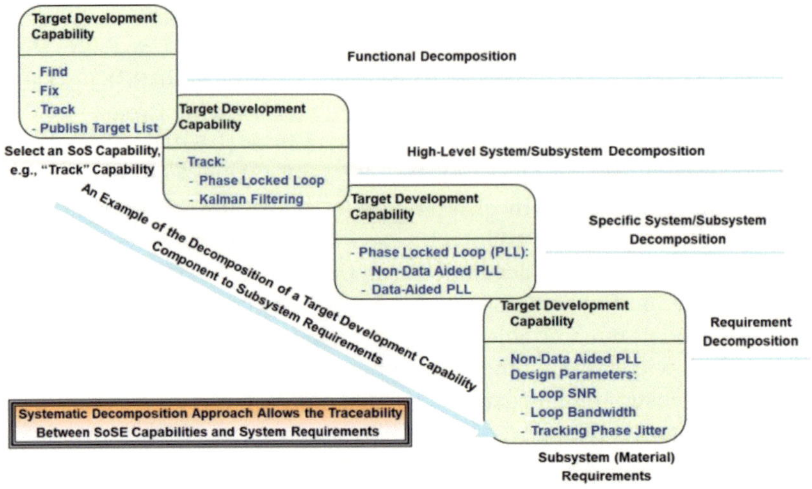

Figure 9.
Example of the functional decomposition of desired capabilities to subsystem requirements for a notional military AOC.

requirements. The "Material" requirements include Facility, system hardware, software, and middleware requirements. The "Non-Material" requirements include D, P, O, P, T, and L.

Figure 8 provides an example of the decomposition of the notional eight key SOS Capability Components into desired capabilities for the notional military air operation center described in **Figure 7**. For this example, the SOSE Capability Component titled "Target Development Capability" component is decomposed into four desired capabilities, namely, Find, Fix, Track, and Publish Target. **Figure 9** provides the "Functional Decomposition" of desired "Find- Fix-Track" Capabilities to "System Requirements" for flowing-down to hardware and software components' requirements. In the example shown in **Figure 9**, the "Target Development Capability" component with the desired "Track" performance can be met by using existing Phase Locked Loop (PLL) and Kalman filtering technology enablers.

4. System architecture design and analysis in SOS environment

This section describes an approach to develop and design a system architecture solution in a SOS environment. Using the proposed approach, the architecture solution will meet customer's requirements for interfacing with existing "as-is" legacy and "to-be" systems. The approach presented here focuses on the use of only OVs and SVs presented in **Figure 4** for capturing the system architecture products, including hardware, software, and middleware components. These components will be captured in a Bill of Material (BOM), which will be used for Rough Order of Magnitude (ROM) costing estimate of the system solution early during the design cycle. **Figure 10** describes the proposed approach for a system architecture design and implementation in a SOS environment. As shown in **Figure 10**, the approach starts with the box labeled "Enterprise Needs" where requirements (or needs) are provided by customer (or users) of the system and/or SOS. Next, the box labeled "Capability Gap Analysis" is performed to determine the "functional gaps" between the required "SOS Enterprise Needs" and existing systems' capabilities. For space applications, the box labeled "Identify Technology Enablers (TEs)" is to conduct a survey of space industry to identify a set of potential TEs that can be used to fill

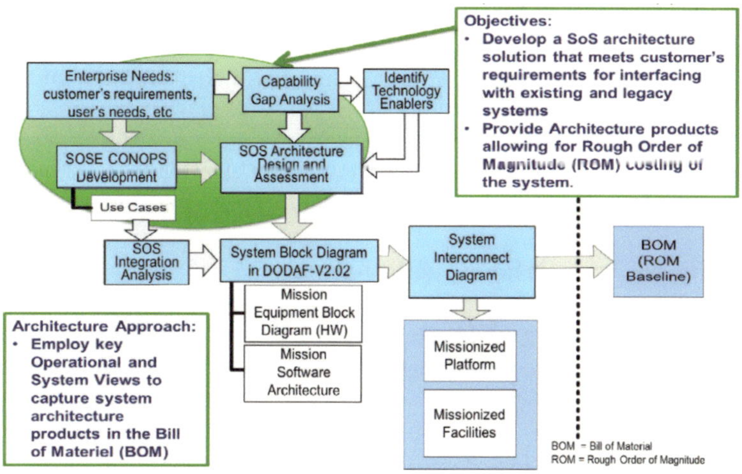

Figure 10.
SOS Architecture design and implementation approach.

the identified functional gaps. At this point, a set of alternative SOS Enterprise Architectures (SOSEA) is derived and assessed to determine the best solution. The box labeled "SOS Architecture Design and Assessment" uses the identified TEs and the SOSEA CONOPS (box labeled "SOSE CONOPS Development"), for the design and select the best system architecture solution from a SOS perspective, namely, a SOS Architecture solution. For traditional "SOS Architecture Design," the approach is based on the "Requirement-Based Architecture Design" approach, which requires a given set of system requirements.[4] For this approach, the architecture trade space is defined by the systems requirements based on a selected set of technology enablers and a corresponding set of specified values for the Measure Of effectiveness (MOEs) as shown in **Table 1**. As an example, for the MOEs, a set of Key Performance Parameters (KPPs) for SATCOM network can be Bit Error Rate (BER), Packet Error Rate (PER), and Packet Loss Rate (PLR).

For long-term customer's needs, the TEs are usually unknown, and the traditional SOS Architecture design approach is no longer applicable, the SOS engineers use "Capability-Based Architecture Design" approach. This approach evolves the "As-Is" architecture to the "End-State" (or "To-Be") architecture solution by going through "Increment" steps. For this approach, the end-state capabilities are derived based on the identified capabilities to fill the "functional gaps" for long-term customer's needs. The increment steps are defined by decomposing the "end-state" capabilities into smaller sets of capabilities that can be addressed by the current TEs at each increment.

The box labeled "System Block Diagram in DODAF-V2.02" uses the SOS Architecture solution derived from the box labeled "SOS Architecture Design and Assessment" and the box labeled "SOS Integration Analysis" to generate a system block diagram using only OVs and SVs (see Section 5.3) with (i) all internal connectivities among subsystems and their hardware, software, and middleware components are identified with actual products' names, and (ii) external SOS connectivities with other systems are identified along with existing Commercial of the Shelf (COTS) interfaces and industry standards. The box labeled "System Interconnect Diagram" captures all required subsystems and their hardware, software, and middleware components and their internal connectivities among

[4] This is also referred to as "Level A" specification. Level A specification is based on a set of selected technology enablers.

themselves. Finally, the box labeled "BOM (ROM Baseline)" is an excel spreadsheet that captures required subsystems and their actual COTS products for all hardware, software, and middleware components and their ROM cost estimate for the overall system.

The following subsections, Sections 4.1, 4.2, 4.3, and 4.4, provide proposed baseline approaches for "Capability Gaps Analysis," "Identify Technology Enablers," "SOS Integration Analysis," and Capabilities Management, respectively.

4.1 Capability gap analysis for capability-based SOS

Figure 11 presents a proposed baseline approach for capability gaps analysis. The flow of this proposed approach is straight forward. The three key features that make this approach unique are:

- Key Feature 1: Conversion of existing system requirements into required current system/SOS capabilities.

- Key Feature 2: Development of a matrix to align existing system requirements with current system/SOS capabilities.

- Key Feature 3: Prioritization of the capability gaps (also referred to as functional gaps) based on mission objectives, performance (e.g., KPPs), technology, and cost requirements.

The next step after the capability gap analysis is the generation of the required capabilities to fill the identified functional gaps. This is also a nontrivial task, as pointed out in Section 3.4, the framework for constructing the capability generation models must be able to generate the required SOS capabilities (to fill the gaps) that are synchronized with the customer's changing needs. For a military enterprise, **Figure 12** proposes an enterprise capability planning framework for generating required capabilities that can be synchronized with changing customer's needs. The salient features associated with this proposed framework are:

- Each of the proposed capabilities generated from this framework is evaluated using Advanced Modeling and Simulation (AM&S) tools to ensure alignment with the identified functional gap and desired operational effects. Example of AM&S tools include SOS AM&S models that can be used to characterize SOS TPMs (e.g., SOS Resilient Capacity, SOS Spectrum Resiliency, etc., see Section 5.1 for more details) for a specified space mission.

- Capability-to-Requirement Conversion and Requirement-to-Capability Alignment processes to ensure system requirements and required customer's/users' capabilities are always synchronized.

- Generation of required capabilities is focused on generating Effects-Based Operation (EBO) capabilities.

- For military applications, generating of required capabilities is based on:

    - DOTLmPF-P trade analysis for best solution: material vs. non-material solutions

    - Total Ownership Cost reduction

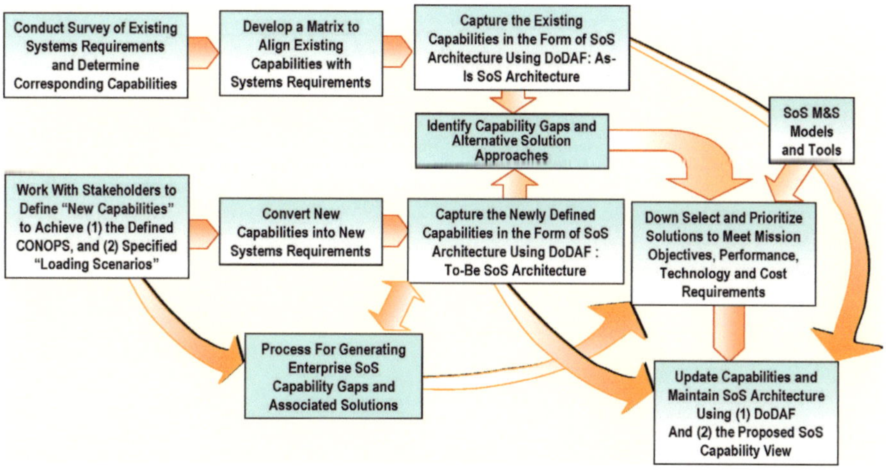

Figure 11.
A capability gap analysis framework for capability-based SOS.

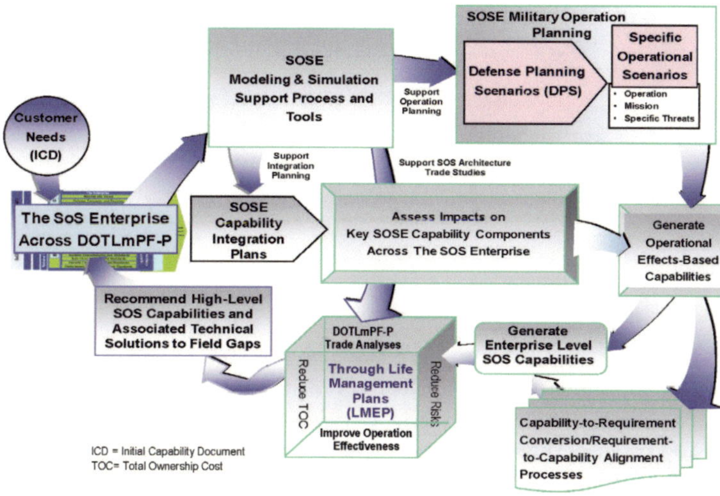

Figure 12.
Proposed framework for generating military SOSE capabilities and associated technical solutions for enterprise planning purpose.

- Improve operation effectiveness (see **Table 1** for measure of operation effectiveness)
- Life Management Plans (LMEP)

The proposed approach focuses on generating the required capabilities using operational EBO approach. **Table 2** illustrates (a) EBO physical action, and (b) EBO kind of effects. For EBO approach, it is critical to define performance measure and develop corresponding advanced modeling and simulation tools to evaluate and assess EBO effects. Note that one can modify the framework presented in **Figure 12** for civilian and commercial enterprises.

4.2 Technical framework for identifying technology enablers

Figure 13 describes a baseline approach for identifying required Technology Enablers (TEs) for a selected SOSEA solution. The approach assumes a notional

(a) EBO Physical Action	
What Is Done	How It Is Done
What	Scale • Force Used • Impact
With What	Scope • Geographic • Operational
	Timing • Speed • Duration • Synchronicity
	Visibility

(b) EBO Kinds of Effects	
Physical	Psychological (Reason/Belief)
Destruction	Chaos/Entropy
Physical Attrition	Foreclosure: • Passive • Active
Chaos/Entropy	Shock
	Psychological Attrition

Table 2.
Description of EBO and associated effects.

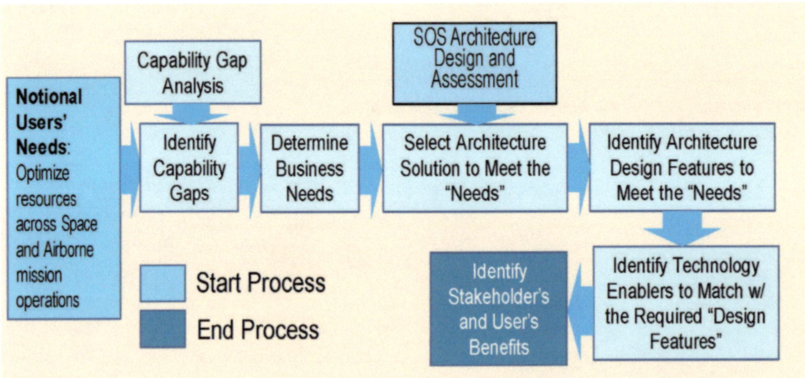

Figure 13.
Technical framework for identifying technology enablers for notional users' needs.

user's needs for space (or airborne) systems. The key step here is the identification of the "Design Features" to meet the Needs. The needs here can be mission needs, or warfighter needs or customers' needs. In this proposed approach, these needs have been translated to "Business Needs." As an example, the business needs for this notional users' needs are ([5], also see Section 5.2):

- Reduce ground system cost
- Improve interoperability

Examples for the selected TEs that meet these business needs are [5]:

- Government of the Shelf (GOTS) multi-mission Satellite Operation Center (SOC) software suite for Telemetry Tracking & Commanding (TT&C) services: Real-time execution component, Mission planning component, Flight dynamic component, etc.
- COTS Satellite Operations (SATOPS) TT&C services
- COTS Open Source Middleware: Active MQ, VMware with Vsphere, etc.

4.3 Integration analysis for capability-based SOS

Figure 14 shows an integration analysis framework for (i) identifying potential SOS integrations problems, and (ii) recommending corresponding solutions for

DOI: http://dx.doi.org/10.5772/intechopen.92674

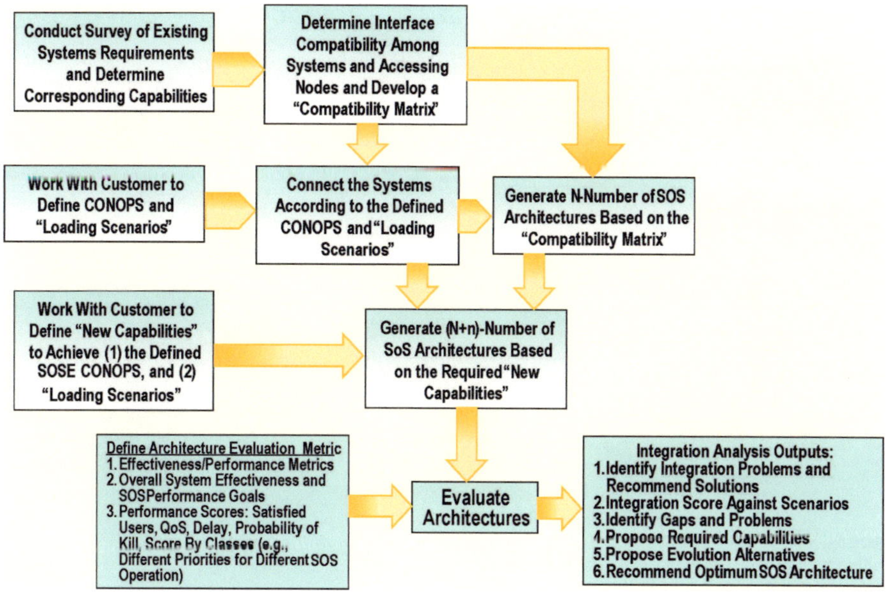

Figure 14.
An integration analysis framework for capability-based SOS.

the identified integration problems. As described in **Figure 14**, the proposed SOS integration analysis allows for incremental architecture solutions discussed earlier and provides the following six outputs:

- Output 1: Identify integration problems and recommend solutions
- Output 2: Integration score against scenarios (or use cases). This score allows the designers understand the effectiveness of the proposed system architecture solution(s) against the threats or "loading" on the systems and network of systems (SOS)
- Output 3: Identify gaps and problems: The gaps here are the internal and external interface gaps in the presence of various threats or loading scenarios
- Output 4: Propose required capability for seamless integration and improve interoperability
- Output 5: Propose evolution alternative(s) in terms of interoperability and affordability.
- Output 6: Recommend optimum SOS Architecture solution in terms of interoperability and affordability

The proposed SOS integration analysis is flexible, agile, and robust and is applicable for any SOS Enterprise because of the following features:

- Compatibility matrix: This matrix is used to track internal and external interfaces ensuring interfaces (i) among internal subsystems and their components are compatible, and (ii) among systems and SOS are compatible. Compatibility matrix is also used to identify the incompatibility among interfaces of any SOS Enterprise.

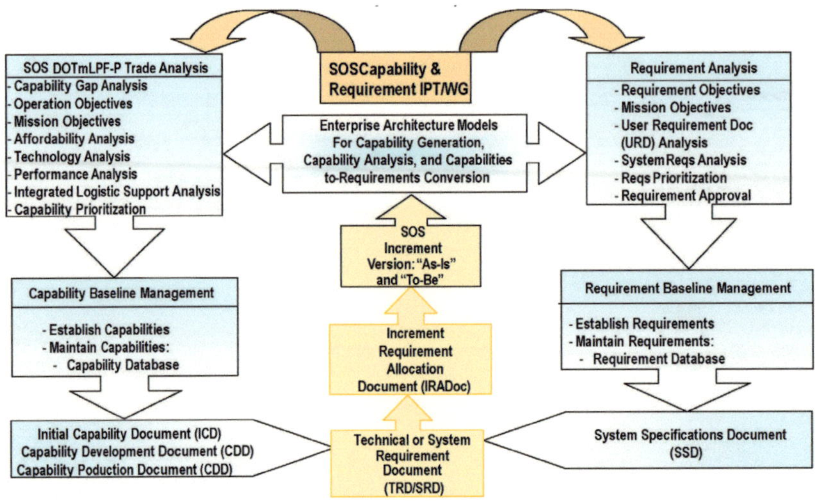

Figure 15.
Capability management framework for a notional.

- Loading scenarios are used in the development of the SOSE CONOPS, where the desired system and SOS loading factors are clearly defined. Loading factors can be the number of users, or data rates, or the number of missile fires, etc.

- Generating alternative SOS Architecture solutions using compatibility matrix. This concept provides a flexible and agile approach for generating practical and implementable SOS Architecture alternatives.

4.4 Capability management framework

A flexible and robust framework to manage systems and SOS capabilities and system requirements for a notional military enterprise is illustrated in **Figure 15**. The proposed framework allows for managing both current/legacy systems requirements and mid-term/long-term systems' capabilities at the enterprise level. The framework ensures capturing system requirements and system capabilities using the existing standard documentation approach and allocating increment requirements for evolving "as-is" to "to-be" systems effectively. Currently, for U.S. military enterprise using capability-based architecture approach, the system capabilities are documented in the Initial Capability Document (ICD), the Capability Development Document (CDD), and the Capability Production Document (CPD) depending on the acquisition phase of the system life cycle [6, 7]. For requirement-based architecture approach, the system requirements are usually documented in Technical Requirement Document (TRD) and/or System Requirements Document (SRD), which are/is derived from System Specification Document (SSD) and/or ICD [7].

5. Examples on notional SOSE Architecture solutions

This section provides three examples related to the design, modeling, simulation, and analysis of a system in SOSE environment. Section 5.1 describes a notional commercial system solution that can be used to augment existing military system for increased resiliency against radio frequency interference threats. Using the architecture design approach presented in this chapter, Sections 5.2 and 5.3 present notional architecture solutions for a typical Satellite Operation Center (SOC) and a notional airborne Intelligence, Surveillance, and Reconnaissance (ISR) platform, respectively.

5.1 A notional SOS solution for increased SOS resiliency

As indicated in Footnote 3, the author has defined and developed two new resiliency metrics addressing Radio Frequency Interference (RFI) problems. The first metric is Resilience Assessment Index Against RFI (RAI-RFI) and the second metric is Spectrum Resiliency Assessment Index (SRAI). Mathematical models for these two metrics can be found in [8, 9]. This section inverts the question addressed in the notional SATCOM SOSE CONOPS described in **Figure 2**. Instead of addressing the question related to the optimum commercial satellite location in space, it addresses the question related to an optimum location for a ground system solution given that the SaOI Node X in space will be defined as one of the three civilian satellites given in the notional SOSE CONOPS presented in **Figure 16**. The SOSEA design issue here focuses on the identification of an existing Military or Civilian Ground Tracking Station (GTS) that can enhance the resiliency of the satellite operations supporting a notional military satellite network described in SOSE CONOPS presented in **Figure 16** [9]. This notional SOSE CONOPS includes 6 military GEO satellites and 8 military GTSs, three civilian satellites, and 15 civilian GTS. Using the RAI-RFI mathematical model presented in [6, 7], **Figure 16** shows a time average heat map of the RAI-RFI metric [6, 7]. The heat map shows the potential optimum GTS locations for improved resiliency [9].

5.2 A notional satellite operation center architecture solution

Figure 17 illustrates an operational view using DODAF OV-5 for a typical SOC operational architecture [5]. The design question here is to upgrade existing SOC using Commercial of the Shelf (COTS) and Government of the Shelf (GOTS) Hardware (HW), Software (SW), and Middleware (MW) TEs. Using the above

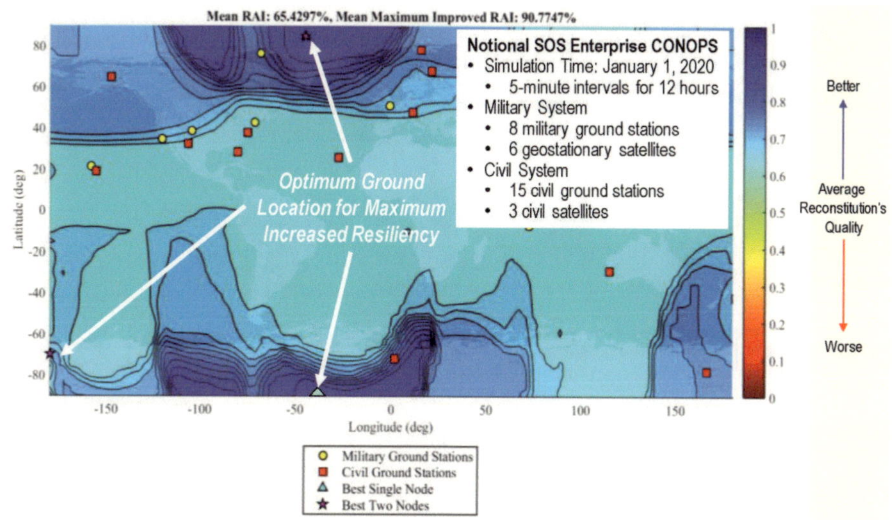

Figure 16.
Example of time average heat map of RAI-RFI metric for a notional military enterprise CONOPS [9].

architecture design approach, one can develop **Table 3** to capture the identified business needs, desired SOC design features (to meet the needs), required TEs, and stake holder and user's benefits [5].

From the required TEs presented in the sixth column of **Table 3**, a notional list of potential TEs that meets the design criteria is presented in **Table 4** [5]. To identify an optimum SOC Architecture solution for the upgrade, the SOS designer needs to conduct a survey to understand the risks associated technology readiness level and market availability of each identified TE and perform an architecture trade study. Using the notional survey results, in Ref. [5], it is shown that the optimum architecture solution for the SOC upgrade is a combination of TE-1, TE-2, TE-4, TE-5, and TE-6.

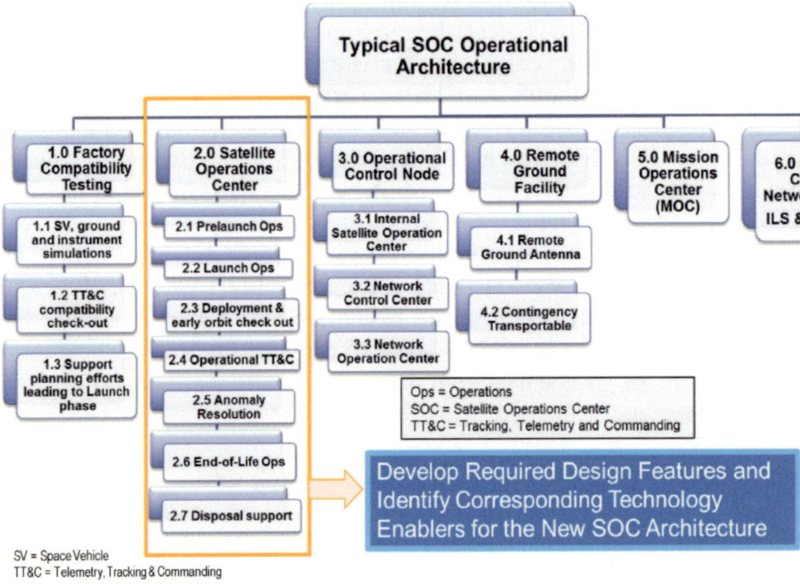

Figure 17.
OV-5: typical SOC operational architecture.

Synchronization the User Needs with "Upgraded" SOC Design Features and Associated TE						
Assumed User Needs	**"Upgraded" SOC Challenge and Gap**	**Business Needs for New SOC**	**"Upgraded" SOC Architecture Design Approach to Meet Business Needs**	**"Upgraded" SOC Architecture Design Feature to Meet Business Needs**	**Required SOC Technology Enablers (TEs)**	**Stakeholder and User Benefit**
Optimize resources across Satellite mission operations	(1) High LCC for modernization of ground C2 system, (2) Contractors "own" the protocol stack which defines the ground SATOPS architecture	Reduce SATOPS ground System Cost and improve Interoperability	Automate Repetitive TT&C Tasks to support multiple missions	Automate Integration with Antenna Scheduling	(1) GOTS MMSOC Software Suite for TT&C Services, (2) GOTS Neptune/ CGA software suite for TT&C Automation	(1) Reduces operation cost, (2) Reduces manpower for operation, (3) Reduces human errors, (4) Reduces interruption of mission, (5) (3) Rapid Delivery without Expensive Integration Costs
				Automate Ground Station Control and Status		
				Automate Data Forwarding for Analysis		
				Automate Command Plan Execution		
				Automate Anomaly Detection		
				Automate Turnkey TT&C System	COTS SATOPS TT&C Services	
			COTS Hardware, Middleware and Software interfaces using MOSA Approach to support multiple missions	Common JIE Compliant Interfaces	Not applicable to our problem. Only Acquiring SOC	(1) Reduces Life Cycle cost, (2) Reduces technology refreshment cost, (3) Increases interoperability, (4) Increases in commonality in Protocol "Stack" for SATOPS ground system
				(1) Virtual AFSCN Receive Network, (2) Outsourcing selected SATOPS received functions and services	(1) COTS MMSOC HW suite, (2) GOTS/ COTS MMSOC Middleware suite, (3) COTS middleware suite	
				(1) Virtual AFSCN Transmit Network, (2) Outsourcing selected SATOPS transmitted functions and services	Not applicable to our problem. Only Acquiring SOC	

Table 3.
Synchronizing user needs with business needs, design features, and technology enablers.

TE No.	TE Name	TE Description	Potential Supplier/KTR
TE-1	GOTS Multi-Mission SOC Software Suite for TT&C Services	Real-Time Execution Component (RTEC): Perform RT execution for satellite bus control	Government Of The Shelf (GOTS) software suite developed by SMC/AD and AFRL using NASA open soure codes
		Mission Planning Component (MPC): Perform mission planning for satellite bus control	
		Flight Dynamics (Orbit Analysis) Component (FDC): Perform flight dynamics for satellite bus control	
		Satellite Engineering Component (SEC or Post-Past Processing Module): Perform satellite bus telemetry acquisition and processing	
		Data Distribution Component (DDC): Perform TT&C satellite bus data distribution AFSCN, Universal Space Network (USN), NASA Integrated Space Network (NISN), Factory and Users	
TE-2	GOTS Neptune/CGA Software Suite for TT&C Automation	User Configurable Extension including: Displays, Databases, Automation, Scripts	GOTS software developed by NRL
		Mission Unique Software including: Command & Telemetry code modifications, hardware drivers	
		Reuse Software Core including: Common C2 function, new capability development, support user CONOPS	
TE-3	COTS SATOPS TT&C Services	Commercial Of The Shelf solution for SATOPS: Space Vehicle (SV) and Ground Station Command and Control (C2) provides TT&C operations support	Universal Space Network (USN), Astrium, Intelsat, Avanti, Kratos ISI, Orbital Systems, etc
TE-4	COTS Multi-Mission SOC HW suite	RT Logic T500GT Gateway Unit: Provides network interfaces to existing Satellite Control Network	GOTS hardware suite developed by SMC/AD and AFRL
		PEGASUS KI-17 Crypto Device: Provides encryption services for satellite telemetry, command link and mission data transfers	
		RT Logic T501 FEP: Performs intelligent front-end telemetry processing, formatting, verification of satellite commands, and control of external devices, such as cryptographic equipment	
		HP Blade Server: Provides servers' services	
TE-5	GOTS Multi-Mission SOC Middleware Suite	GMSEC (Goddard Mission Services Evolution Center) Message Bus Hosted on Open Source COTS Middleware	NASA
TE-6	COTS Open Source Middleware	ActiveMQ open source message broker written in Java	Open Source Code
		Vmware with vSphere	Vmware
		Vmware with vSphere with Operations Management	
TE-7	COTS Middleware Suite	COTS Message Bus Hosted on Open Source COTS Middleware	KTR #1 (TBD)
		Open Source COTS Middleware: ActiveMQ open source message broker written in Java	Open Source Code
		Open Source COTS Middleware: Vmware with vSphere, Vmware with vSphere with Operations Management	Vmware
TE-8	COTS Software Suite for TT&C Automation	Automated SV and Ground Station C2 provides automated TT&C operations developed by Braxton's ControlPointTM	Braxton's ControlPoint™ provides Automated TT&C Operations

Table 4.
A notional list of technology enablers and associated suppliers.

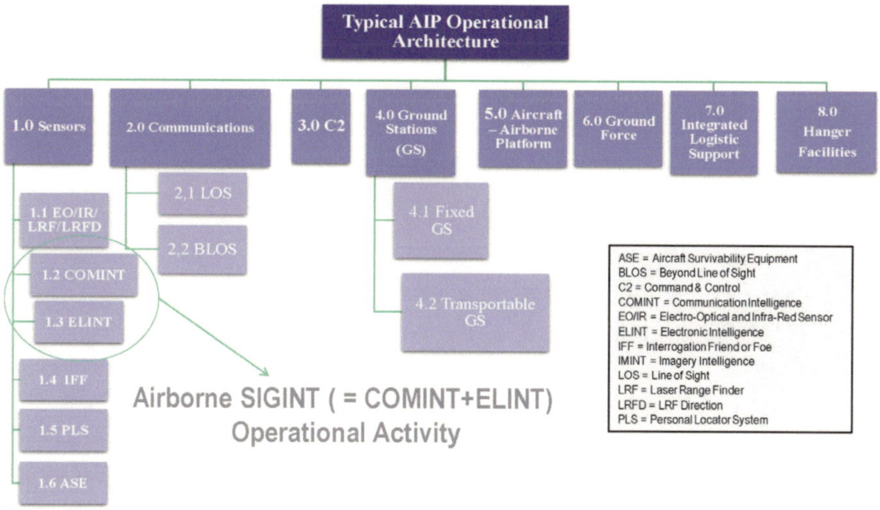

Figure 18.
OV-5: Typical AIP SOS operational activity.

5.3 A notional airborne ISR platform architecture solution

This section presents how a DODAF views presented in **Figure 4** can be used to capture a typical Airborne ISR Platform (AIP) architecture solution operating in a notional SOS environment. The notional SOS environment considered here includes: (i) Military user nodes that can be on ground or surface or air, and

(ii) Commercial ground nodes that can be a ground broadcast news or a commercial mobile user. **Figure 18** describes a DODAF OV-5 for typical AIP SOS operational activity. **Figure 19** illustrates a DODAF OV-2 for typical AIP SOS operational node connectivity. **Figure 20** illustrates a DODAF SV-1 for typical AIP SOS interface in the notional SOS environment.

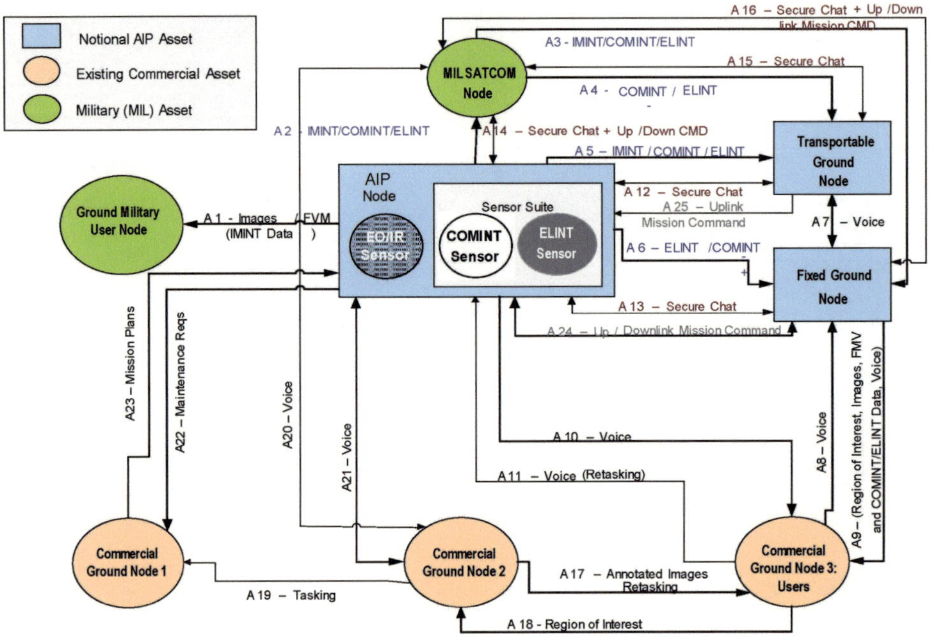

Figure 19.
OV-2: Typical AIP SOS operational node connectivity.

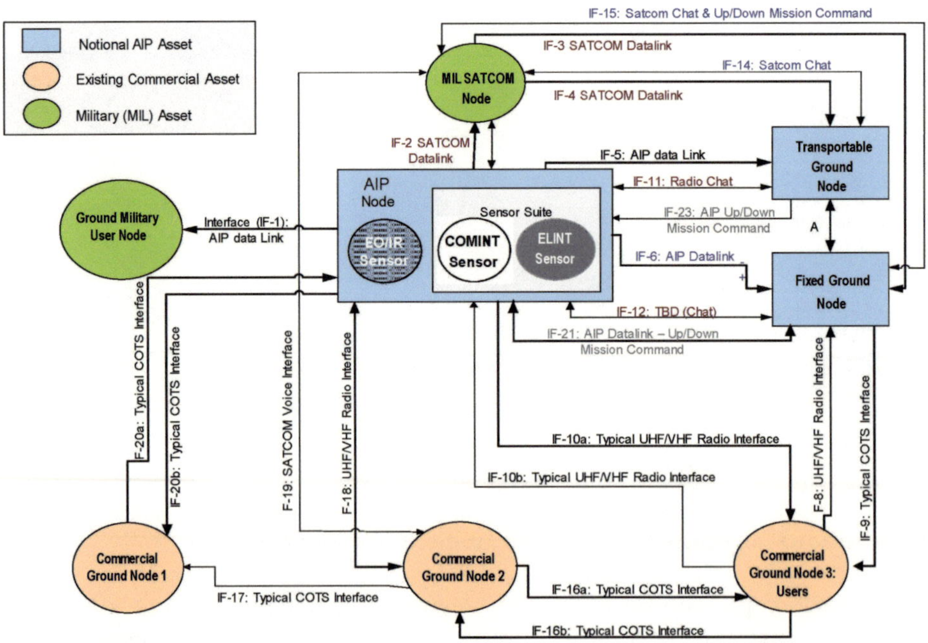

Figure 20.
SV-1: Typical AIP SOS interface in a notional SOS environment.

6. Conclusion

This chapter describes SOSE, SOSE CONOPS, SOSEA alternative solutions, SOSEA CONOPS assessment, and SOSEA design approach through examples using typical and/or notional space and airborne systems in a typical SOS operational environment. The SOSE environment considered is a combination of military, civilian, and commercial operational environments. As pointed out throughout the chapter, examples for military applications can be tailored for civilian and commercial applications and vice versa. Similarly, examples for airborne systems can also be tailored for space systems considering transmission delay for satellite systems is much longer than airborne systems. The SOSEA approach proposed in this chapter has been used by the author for the design, modeling, simulation, and analysis of space and airborne systems for actual military and commercial programs.

Acknowledgements

The author would like to express his appreciation to Prof. Charles Lee and Prof. Alfonso Agnew, Chair, Department of Mathematics at California State University, Fullerton.

Conflict of interest

The preparation of this chapter was not funded by the Aerospace Corporation, and it was done by the author using his own time and resources; thus it does not represent the Aerospace Corporation's view on SOS, SOSE and SOSE Architecture.

Notes/thanks/other declarations

The author wishes to thank his wife, Thu-Hang Nguyen, for her enormous patience and boundless support during the preparation of this chapter.

Author details

Tien M. Nguyen[1,2]

1 California State University in Fullerton, California, USA

2 The Aerospace Corporation, United States

*Address all correspondence to: tnguyen57@aol.com

References

[1] Dahmann J, Lane JA, Rebovich G Jr, Baldwin KJ. A Model of Systems Engineering in a System of Systems Context. MITRE Corporation; 2008. Available from: https://www.researchgate.net/publication/228576609_A_model_of_systems_engineering_in_a_system_of_systems_context

[2] Henshaw M, Dahmann J, Lawson B. Systems of systems. In: System Engineering Book of Knowledge (SEBoK); 2019. Available from: https://www.sebokwiki.org/wiki/Systems_of_Systems_(SoS)

[3] Department of Defense Architecture Framework. DODAF Version 2.02; 2011. Available from: https://dodcio.defense.gov/Portals/0/Documents/DODAF/DoDAF_v2-02_web.pdf

[4] Burch R. A method for calculation of the resilience of a space system. In: IEEE Military Communications Conference Proceedings; 2013

[5] Nguyen TM, Andy Guillen. Acquisition war-gaming applications for acquiring future ground satellite control systems. Poster Presentation. 2017 Ground System Architecture Workshop (GSAW). The Aerospace Corporation; 2017

[6] DOD Instruction 5000.02; 2013. Available from: https://www.dau.edu/guidebooks/Shared%20Documents%20HTML/DoDI%205000.02.aspx

[7] Nguyen TM, Guillen AT, Matsunaga SS.A resilient program technical baseline framework for future space systems. Vol. 9469. Invited Paper. Pham KD, Chen G, editors. In: 2015 SPIE Conference Proceedings, Sensors and Systems for Space Applications VIII; 2015

[8] Nguyen TM, Lee CH, Freeze T, Guillen AT. Systems-of-Systems enterprise architecture CONOPS assessment approach and preliminary results. In: Chen G, Pham KD, editors. Invited Paper. Vol. 9469. To be published in: 2020 SPIE Conference Proceedings, Sensors and Systems for Space Applications VIII; 2020

[9] Freeze T, Nguyen TM, Lee C. SOS Enterprise CONOPS assessment decision support tools. In: Nguyen TM, editor. SOS Design—Engineering, Modeling, Simulation and Analysis; Intech Open Publisher; 2020 [Unpublished]

Chapter 5

System-of-Systems Enterprise CONOPS Assessment Decision Support Tools

Thomas O. Freeze, Tien M. Nguyen and Charles H. Lee

Abstract

This chapter discusses the implementation of System-of-Systems Enterprise Architecture (SOSEA) CONOPS assessment framework and models in Matlab, and presents preliminary results concerning SOSEA resiliency in the presence of a notional Radio Frequency Interference (RFI) scenario. The chapter provides an overview of the SOSEA CONOPS Assessment Framework, and discusses related SOS Resiliency Models including Resilient Assessment Index Against RFI (RAI-RFI), Spectrum Resiliency Assessment Index (SRAI), and Resilient Capacity (RC).

Keywords: SOS enterprise architecture, CONOPS, resilient capacity, resilience assessment index against radio frequency interference, spectrum resiliency assessment index, satellite communication, satellite operation

1. Background and introduction

In 2011, The Department of Defense (DOD) established a formalized concept of resilience for space systems and Systems-of-Systems (SOS) [1]. They define resilience as the ability of an architecture to support the functions necessary for mission success in spite of hostile action or adverse conditions. Similarly, in Enhancing Space Resilience Through Non-Material Means McLeod, et al. define resilience as "an attribute of a system that generally indicates its ability to maintain critical operations in the face of adverse disruptions" [2]. However, they acknowledge that there is much room for variation to this definition depending on circumstances and priorities. Gregory Edlund splits resilience evaluations into two broad categories of either being analytic or deterministic [3]. For Edlund, analytic models are geared more towards attempting to measure or score a system's resilience. Deterministic models are focused more on characterizing a system's breaking points, which Edlund argues may be more useful in practice. This chapter addresses the modeling of SOS Enterprise Resiliency and its metrics.

The U.S. Military uses communications satellite systems to facilitate beyond line-of sight communications. For decades, the satellite space has been largely uncontested. However, the orbital space around the Earth has become more congested as technology advances. Such advances make the space environment more accessible to various organizations. Third party un-intentional interferences with the military satellite communication system are a growing and serious threat as more and more government, commercial, and civilian entities enter the orbital space environment. This chapter discusses three models to analyze different resiliency aspects of the military's satellite space system against the threats caused by third party RFI with a focus on unintentional interferences. The first model is the Resilience Assessment Index Against RFI (RAI-RFI) that will be used to assess the robustness and reconstitution of a SOS [4–7]. The Second model is the Spectrum Resilience Assessment Index (SRAI) [4, 6–8], which is an expansion of the RAI-RFI. By adding spectrum analysis to the RAI-RFI model, the amount of time that a communication link can access its allocated frequency band can be measured in the presence of disruptive events, such as RFI. Various communication technologies can then be compared to identify the best technology for enhanced spectrum resilience. The third model is the Resilient Capacity (RC) model, which assesses a SOS ability against RFI threats [4–8]. Additionally, the RC score will attempt to be improved when RFI causes disconnections by augmenting the military system with a pre-existing commercial or civilian satellite.

This chapter presents the work done in 2018 by CSUF graduate student team with a focus on the Matlab implementation of RAI-RFI, SRAI and RC models. The chapter organizes as follows: (i) Section 2 provides definition of SOS resiliency and its metrics for evaluation the resiliency; (ii) Section 3 discusses the differences between SOS Enterprise (SOSE) CONOPS and SOSE Architecture (SOSEA) – Section 3 also provides description a notional Satellite Operation (SATOP) SOSE CONOPS and Satellite Communication (SATCOM) SOSE CONOPS; (iii) Section 4 provides an overview of the proposed SOSEA CONOPS assessment approach, including framework and associated models; (iv) Section 5 presents an implementation approach of the framework and models in Matlab commercial-of-the-shelf software; (v) Section 6 discusses SOSE CONOPS Modeling in Matlab; (vi) Section 7 addresses the SOSE RAI-RFI Modeling in Matlab and simulation results; (vii) Section 8 describes SRAI model implementation and simulation results in Matlab; (viii) Section 9 discuss RC Model implementation and simulation results in Matlab; and (viii) Section 10 concludes the chapter with a discussion on the preliminary results and way-forward.

2. Definition of SOS resiliency and its metrics

In this chapter, the metric "Resiliency" is defined in the context of RFI threats, i.e., Resiliency against RFI threats. The RFI threats include both Friendly and Un-friendly RFI threats. The chapter focuses on the following three Resiliency metrics [4]:

- Resilience Assessment Index Against Radio Frequency Interference (RAI-RFI): This is a newly proposed "Robustness-and-Reconstitution" metric that calculates the probability of a ground tracking system or a satellite communication system is being disrupted by RFI events and its ability to reduce RFI by re-routing the desired signal to avoid RFI threats. RAI-RFI provides a measure of SOSE robustness and quality of reconstitution.

- Spectrum Resiliency Assessment Index (SRAI): SRAI is also a newly proposed "Avoidance-Robustness-and-Reconstitution Metric", which derived from the U. K. Ministry of Defense (MOD) that defines the ability of a system that can access the spectrum and be able to response to a disruptive event. SRAI is a metric that is calculated based on the probability that a system can access to its allocated RF frequency band in the presence of un-friendly and/or friendly-RFI threats.

- Resilient Capacity (RC): This metric is derived from the "DOD Definition of Resilience" focusing on Avoidance, Robustness, Reconstitution, and Recovery

DOI: http://dx.doi.org/10.5772/intechopen.94123

RC is defined as the SOSE's probability that two arbitrary nodes within a SOSE network can communicate with each other amidst RFI adversity. It is a function of Avoidance, Robustness, Recovery and Reconstitution. Nodes can be a ground tracking node on the ground or a satellite node in space

3. SOS enterprise CONOPS vs. SOSEA

The differences between SOSE CONOPS and SOSEA are illustrated in **Figure 1**. The detailed description of this figure can be found in [5].

3.1 Notional satellite operation SOSE CONOPS

Figure 2 illustrates a notional SATOP SOSE CONOPS that will be used for the development of the SOS tools presented in this chapter. The detailed description of this figure can also be found in [5].

3.2 Notional satellite communication SOSE CONOPS

A notional SATCOM SOSE CONOPS can be found in [5]. For the purpose of the SOS tool development, **Figure 3** shows the notional SATCOM SOSE CONOPS used for the demonstration of the SOSE tools discussed in this chapter. Note that the communication satellite RFI Node 4 and exo-atmospheric Jammer Node 1 show a potential un-intentional and intentional RFI sources, respectively, that can disrupt

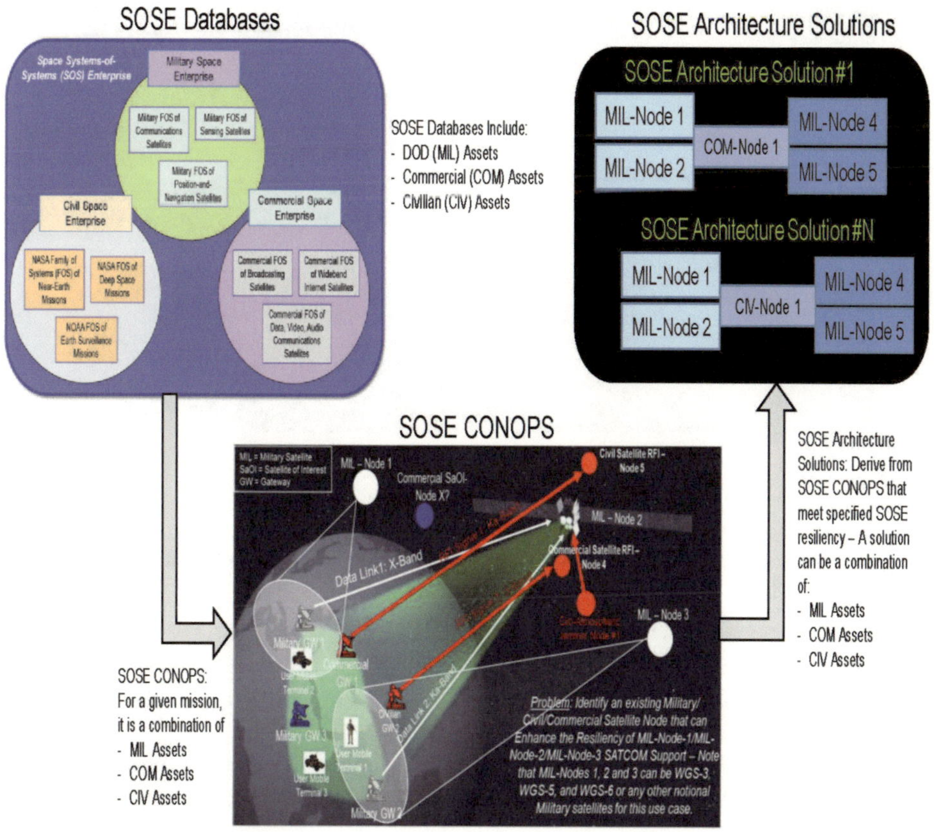

Figure 1.
Overview of SOSE CONOPS and SOSEA solutions [5].

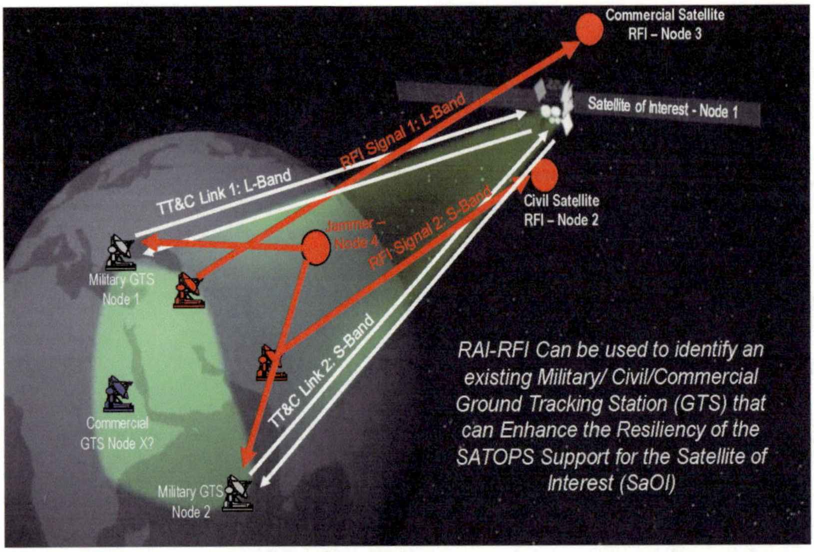

Figure 2.
Notional SATOP SOS Enterprise CONOPS [5].

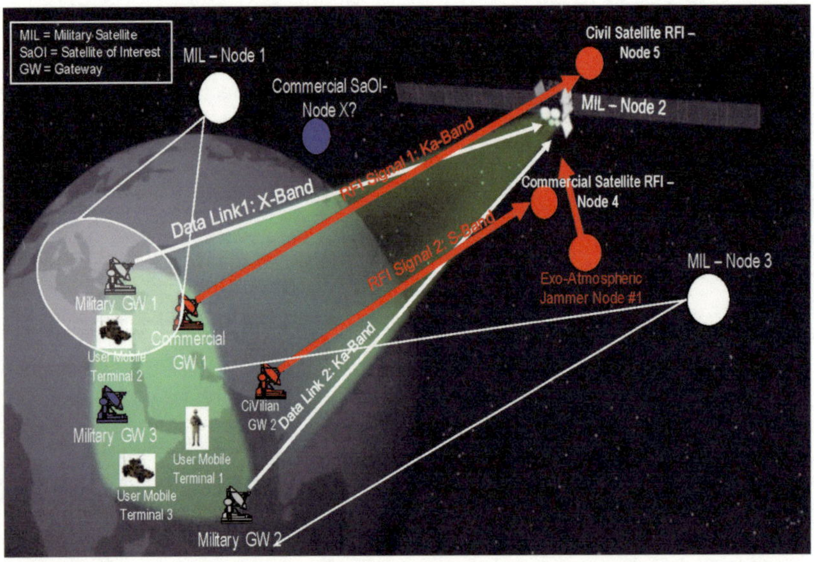

Figure 3.
Notional SATCOM SOS Enterprise CONOPS [5].

the communications links of interest. The simulation results shown in Sections 7, 8 and 9 assumed the intentional RFI Node 1 is off.

4. SOS enterprise architecture CONOPS assessment approach

To address the SOSE Resiliency problem, this section addresses SOSEA CONOPS assessment approach (as shown in **Figure 4**) with the following key SOSE CONOPS framework features:

- SOSEA with three distinct space enterprises consisting of Military Space Enterprise, Commercial Space Enterprise and Civil Space Enterprise.

- Databases include military, commercial and civil satellites and ground systems.

- Four Key SOSE CONOPS Assessment Metrics for measuring the space enterprise performance including Communication Link Margin (focus of this chapter), Communication Link Availability (focus of this chapter), System Availability (not cover in this chapter) and Network Availability (not cover in this chapter)

- Three Key Resiliency Metrics for measuring "Spectrum Resiliency" against RFI threats: RAI, SRAI, and RC.

And, the key SOSE mathematical and simulation models' features are:

- RAI Model: Generates a "Heat-Map" to show areas impacted by RFI threats and associated reconstitution's quality.

- SRAI Model: Generates a "Heat-Map" to show the likelihood that a communication system can access to the allocated frequency-band in the presence of RFI events

- RC Model: Generates SOSE Communication Link Margin and Link Availability for the "areas identified by RAI and SRAI" models.

- System Recovery Time Model: Estimates system recovery time from RFI threats.

Note that the SOSE System Availability and Network Availability metrics are not covered in this chapter. **Figure 4** presents the proposed SOSE framework and

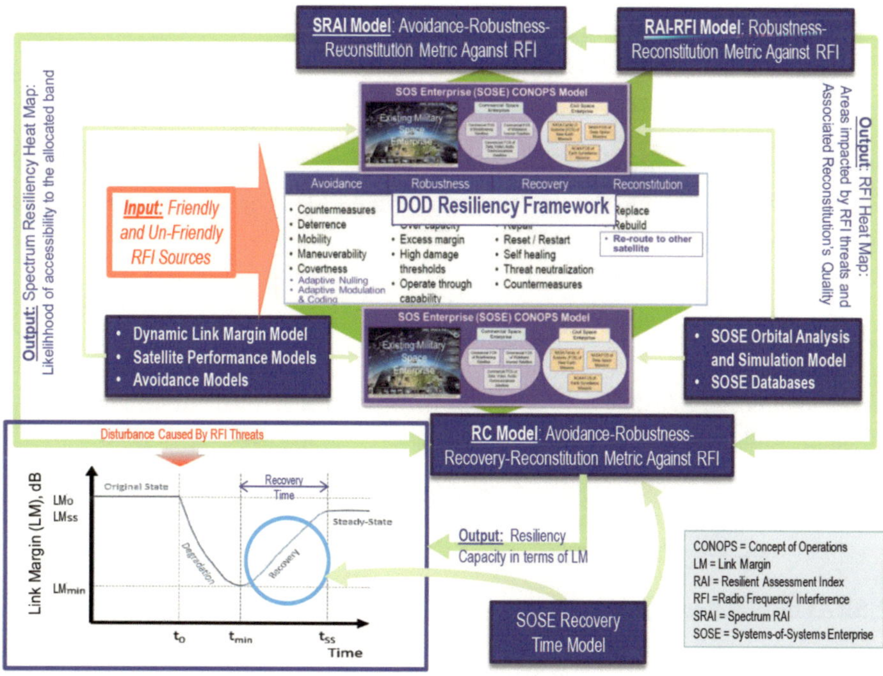

Figure 4.
Integrated SOS enterprise framework and models.

models to be implemented in Matlab for the evaluation of SOS metrics, including RAI, SRAI and RC.

4.1 RAI-RFI model description

RAI-RFI model can be used to select an optimum placement of a new ground station within existing SOSEA to reconstitute ground-station-to-satellite links when adverse RFI events occur. The model can also be used to select the best space and ground network nodes within existing SOSEA when adverse RFI events occur.

Let N_{GS} be the number of Ground Stations (GS) in our SOSE and N_{Sat} be the number of satellites in our SOSE. Then we define p_{ij} at time t under ideal conditions (no RFI) as follows [4, 6]:

$$p_{ij}(t) = \begin{cases} 0, \textit{if GS has no link with Satellite j at time t in ideal case} \\ 1, \textit{if GS has a link with Satellite j at time t in ideal case} \end{cases} \quad (1)$$

for $i = 1, \dots, N_{GS}$ and $j = 1, \dots, N_{Sat}$.

Let P(t) be:

$$P(t) = \sum_{i=1}^{N_{GS}} \sum_{j=1}^{N_{Sat}} p_{ij}(t) \quad (2)$$

We define $\tilde{P}$ at a given time t as follows:

$$\tilde{P}^X(t) = \sum_{i=1}^{N_{GS}} \sum_{j=1}^{N_{Sat}} \tilde{p}_{ij}^X(t) \quad (3)$$

Where

$$\tilde{p}_{ij}(t) = \begin{cases} 0, \textit{if the set of node X does not reconstitute the link} \\ 1, \textit{if node x reconstitutes the link} \end{cases} \quad (4)$$

Note that $\tilde{p}_{ij}(t)$ is only eligible to be 1 if both $p_{ij}(t) = 1$ (a link exists in ideal conditions) and $\hat{p}_{ij}(t) = 1$ (the link is down to due RFI). The RAI-RFI robustness metric is given as [4, 6]:

$$R_{RAI}(t) = \frac{P(t) - \hat{P}(t)}{P(t)} \quad (5)$$

The augmented RAI-RFI metric which incorporates reconstitution via hypothetical ground stations is given as [6]:

$$\tilde{R}_{RAI}(t) = \frac{(P(t) - \hat{P}(t)) + \tilde{P}(t)}{P(t)} \quad (6)$$

4.2 SRAI model description

As mentioned above, the Spectrum Resiliency is the ability of systems that access the spectrum to respond to a disruptive event such as RFI [4]. On the other hand, Section 4.1 shows that RAI is a metric describing if the systems can contact each other. The SRAI metric weights in the band sharing when a contact is made, and it is defined as follow:

$$\mathbf{SRAI} = \mathbf{SRAI_0} + \mathbf{R_{RAI}} \tag{7}$$

where $\mathbf{R_{RAI}}$ is given by Eq. (5), and $\mathbf{SRAI_0}$ can be calculated using a simplified mathematical model presented in **Figure 4**.2 [4, 8]. This simplified model is derived for two popular multiple access techniques, namely, Frequency Division Multiple Access (FDMA) and Time Division Multiple Access (TDMA). **Figure 5(a)** describes the simplified model for calculating$\mathbf{SRAI_0}$, where the SAS Index (SASI) can be calculated using a mathematical model shown in **Figure 5(b)**. Sample calculations of notional SRAI metric for FDMA, TDMA and Code Division Multiple Access (CDMA) can be found in [8].

4.3 RC model description

This subsection describes the RC model derived from the U.S. DOD definition for resiliency [1, 4]. In SOSE context, RC is defined as the SOSE's probability that two arbitrary ground nodes can communicate with each other amidst adversity. It is a function of the following four metrics [1, 4]:

- Avoidance (R_{AV}) is the probability a threat can be avoided or prevented altogether
- Robustness (R_{RO}) is the probability two arbitrary nodes can communicate amidst degradation (i.e. radio frequency links lost due to increased RFI)
- Recovery (R_{RV}) is the probability two arbitrary nodes can communicate when links survive via band flipping (in presence of RFI)

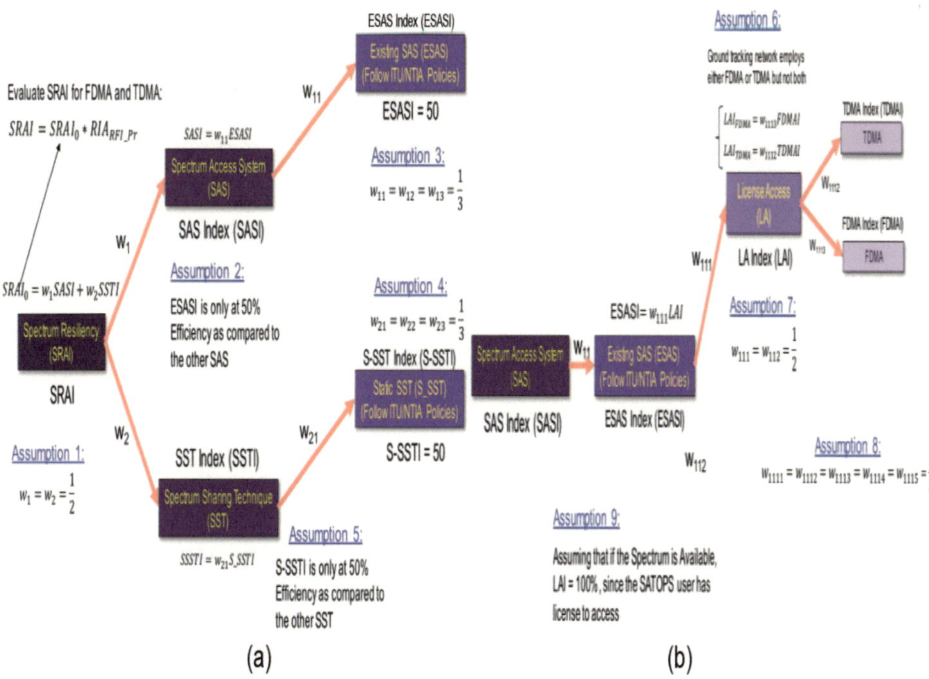

Figure 5.
Simplified SRAI model for FDMA and TDMA. (a) Calculation of SRAIo. (b) Calculation of SAS Index (SASI).

- Reconstitution (R_{RC}) is how likely the system can be re-established to full operational capacity while using 3rd party satellite support.

Figure 6 describes a simplified RC behavior modeling in a SOSE environment. A simple mathematical model can be developed to characterize the RC behavior, and it is given below [1, 4]:

$$R = R_{AV} + (1 - R_{AV})R_{RO} + (1 - R_{AV})(1 - R_{RO})R_{RV} + (1 - R_{AV})(1 - R_{RO})(1 - R_{RV})R_{RC} \quad (8)$$

Note that the RAI model encompasses the "Avoidance" and "Reconstitution" resiliency features, since it is used to select (i) an optimum placement of a new ground station within existing SOSE architecture to "Reconstitute" ground-station-to-satellite links when adverse RFI events occur, and (ii) the best space and ground network nodes within existing SOSE architecture when adverse RFI events occur (i.e., RFI avoidance).

5. Implementation approach: decision support tool

The Graduate student team's approach is to build mathematical models and develop numerical algorithms from scratch. The team implements models using MATLAB without any other Commercial of the Shelf (COTS) software, freeware (e.g., STK, SOAP, NAIF, etc.) to avoid licensing and interfaces. The team collected and maintained a database of ground and space systems including:

- Module-base extension
    - Trajectory propagator (fundamental, core, no black-box software, flexibility, scenario-driven)
    - Resilience calculation (modeling, assessing, etc.)
    - Signal processing (near future, testing advanced techniques)
    - Add-on/Future projects
- Document technical reports, codes, and user manuals (easy to pass on to future students).

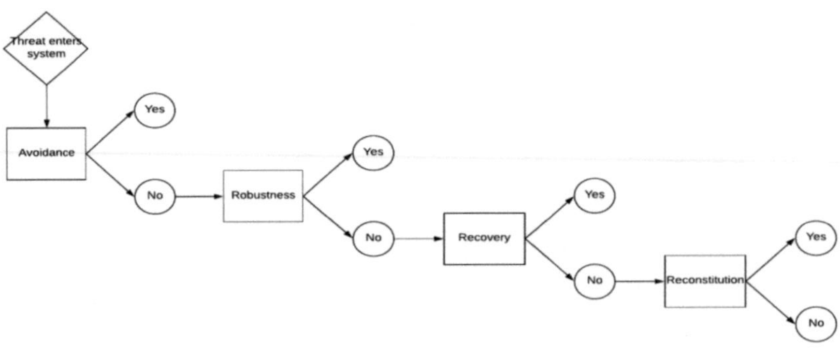

Figure 6.
RC behavior modeling in a SOSE environment.

6. SOSE CONOPS modeling in Matlab

SOSE CONOPS modeling in Matlab includes two key activities, namely, Data collection and input into model simulation (Section 6.1) and orbital dynamic modeling (Section 6.2).

6.1 Data collection and input into simulation

In our proof of concept simulations, we use publicly available data on satellites and ground stations from [6]:

- CelesTrak
- CCSDS
- World Meteorological Organization
- NASA Near Earth Network, Deep Space Network, and TDRSS Network
- Air Force Satellite Control Network (AFSCN)
- NOAA, etc.

The data are then parsed the two-line elements for satellites as seen in **Figure 7** and placed them into our orbital model. Ground Systems were added to the simulation by latitude, Longitude and Height and converted into (x,y,z) tuples used in the Earth Centered Inertial (ECI) model. Once the satellites and ground systems were added to the system their positions are updated using a Kepler propagator [6].

6.2 Orbital dynamic modeling and simulation results

In our simulations we use a simplified "Dynamic Link Margin" (DLM) model [4]. The simplified DLM model assumes the following [4, 6–8]:

- An Approximation of the link budget model with simplified RFI and signal strength degradation model
- A fixed cross-polarization isolation loss
- Zero recovery time

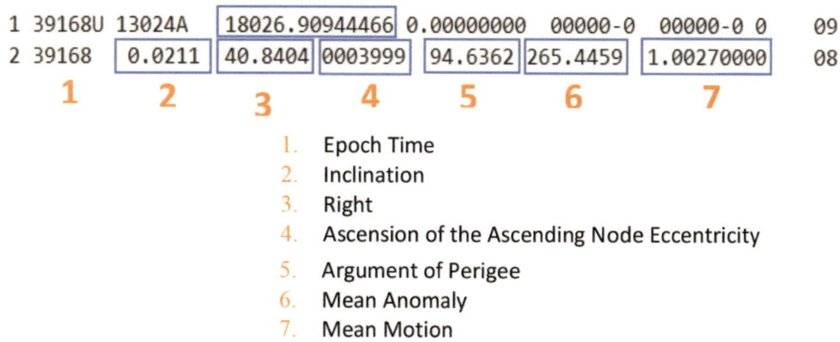

Figure 7.
Two-line elements for satellites [6].

Our first step is to make sure the two systems are in view of each other. **Figure 8** Shows the calculation for ground to satellite and associated diagram of the geometry. **Figure 9** shows the calculated area of coverage for a beam cone along with its resulting coverage.

Once, we calculate that the satellite and ground station have each other in Field-of-View (FoV), We calculate our link margin using our link margin model which factors in antenna geometry and each satellites/ground stations unique parameters.

The In-View and Footprint model implemented in Matlab are shown in **Figure 10** [6]. The model assumed that:

- Minimum elevation angle of ground station is known and is constant
- Satellite beams are circular or elliptical
- Half-power beam width of satellite is known and is constant

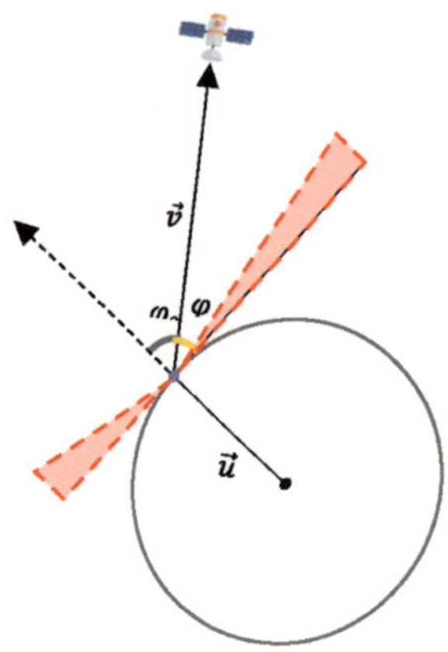

Construction of In-view Line:

- $\vec{u} = \frac{GS\ Position}{\|GS\ Position\|}$, $\vec{v} = \frac{Sat\ Position - GS\ Position}{\|Sat\ Position - GS\ Position\|}$
- We get the co-elevation angle,
  $$\varphi_0 = \cos^{-1}(\vec{u} \cdot \vec{v}) \quad (9)$$
- Satellite elevation angle,
  $$\varphi = 90^{\circ} - \varphi_0 \quad (10)$$
- If $\varphi \geq$ minimum elevation angle, draw line from satellite to ground station

Figure 8.
Calculation for ground to satellite.

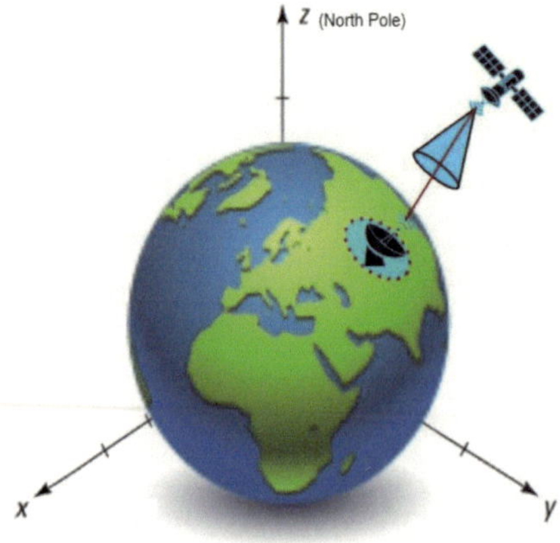

Construction of Beam Cone and Area of Coverage:

- Let θ_{NS} and θ_{EW} be the beam angles in the North-South and East-West directions respectively. The beam angle is denoted by the "Blue" beam cone comes from the satellite
- Construct N many points on an ellipse on the yz-plane that satisfy:
  $$\frac{y^2}{\tan^2(\theta_{EW})} + \frac{z^2}{\tan^2(\theta_{NS})} = 1$$
- Project every point in the x-direction a distance of 1 unit
- Create vectors from origin to each point.

Figure 9.
Calculation for area of coverage for a beam cone [6].

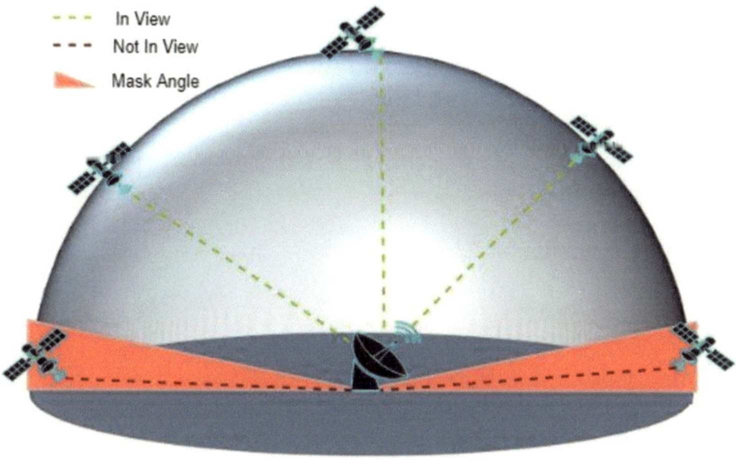

Figure 10.
In-view and footprint model implemented in Matlab [6].

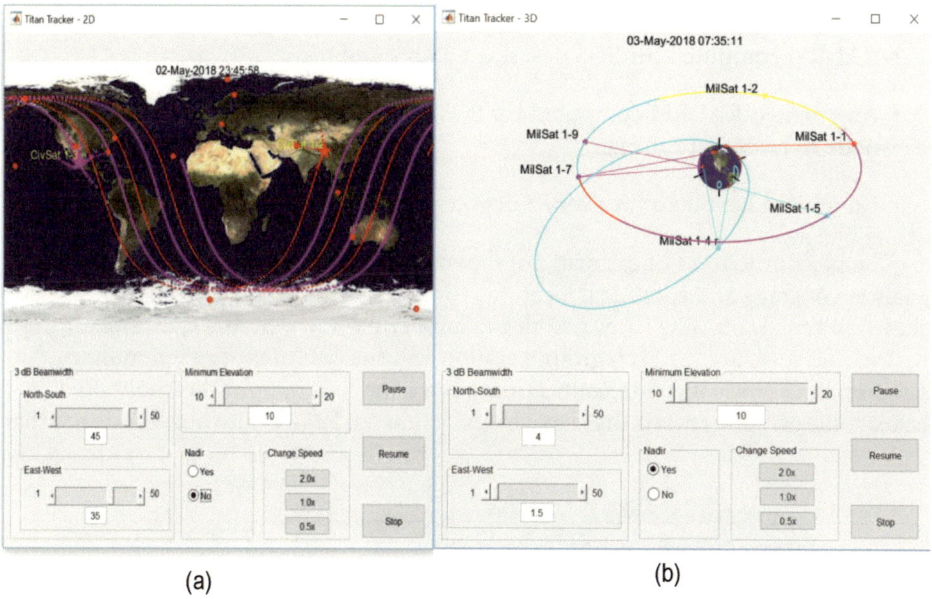

Figure 11.
Matlab simulation results of a notional SOSE scenario [6].

- "Aim point" is Nadir if satellite antenna is not steerable, and towards ground station if satellite antenna is steerable
- A satellite is In-view if its elevation is greater than its minimum elevation angle.

Figure 11 demonstrates the Matlab Simulation Results of a Notional SOSE Model [6]. **Figure 11(a)** shows the 2-D view of the simulation results. **Figure 11(b)** shows the 3-D view.

7. RAI-RFI implementation in Matlab and simulation results

The RAI-RFI model described in Section 4.1 is implemented in Matlab. The following notional SOSE scenario was implemented in the model for demonstration purpose [7].

- Simulation duration: January 1, 2020
    - 5-minute intervals for 12 hours
- Notional Military System
    - 8 military ground stations
    - 6 geostationary satellites
- Notional Civil System
    - 15 civil ground stations
    - 3 civil satellites
- RAI-RFI computed for SOS described above and
- Augmented RAI-RFI computed for SOS with a single hypothetical additional node to create a heat map
- Computed heat map for every 5 degrees of latitude and longitude

The plots of RAI-RFI heat map are shown in **Figure 12** [7]. Where we can analyze coverage and spot ideal locations for adding in new ground stations as shown in the “Dark Blue” areas. Link availability is shown on the right side of the figure. The rows are for each ground station and the columns for each military satellite. Blue boxes indicate connections above the link margin threshold. Red boxes indicate connections that have been lost due to RFI. White boxes indicate that

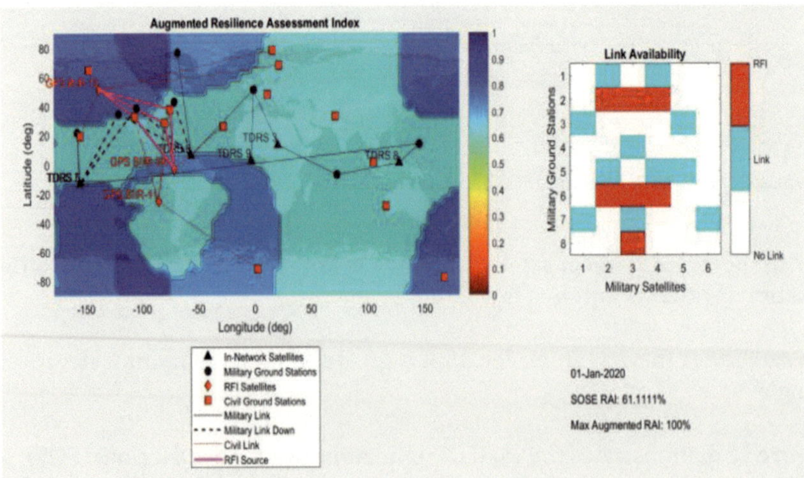

Figure 12.
RAI-RFI plot: time-averaged heat map [7].

no connection is possible between the ground station and satellite which is usually due to geometry. At each time interval the link availability chart is recalculated.

8. SRAI implementation in Matlab and simulation results

As discussed in Section 4.2, the Spectrum Resiliency Assessment Index (SRAI) is an add on to RAI that considers issues of spectrum access and usage [8]. This metric augments our tools by allowing us to consider spectrum access policy and spectrum access interference use cases. Existing SATOPS are conducted using FDMA and TDMA. We want to demonstrate that using CDMA we could increase SRAI, and hence increase spectrum resiliency. **Table 1** summarizes our reasoning for the calculation of SRAI index for FDMA, TDMA and CDMA [4].

The intent is of this section is to demonstrate the tool capabilities, hence the details of the simulation scenario is not provided here. But it should be noted that the RFI sources considered in this section are the interferences from friendly satellites and ground stations shown in **Figure 3** [5], and the amount of interference power is calculated assuming that the interferer satellites used omni antenna and transmitted power was extracted from our database. For the SATOPS scenarios shown in **Table 1**, the results show that FDMA is only 5% efficiency as compared to CDMA and TDMA is 20% efficiency as compared to CDMA. For the same notional SOS scenario presented in Section 6, using the results presented in **Table 1** along with Eq. (7) and **Figure 5**, the following 3 figures illustrate how different access protocols can affect the SRAI results:

- **Figure 13**: SRAI for FDMA model described in Section 4.2. For the notional SOS scenario described in Section 7, the simulation results show that the FDMA allowed very limited spectrum access to share a frequency at once. The SRAI index is less than 0.1.

SATOPS	FDMA	TDMA	CDMA
Allocated Bandwidth	25 MHz	25 MHz	25 MHz
Frequency reuse	2	2	1
Required channel BW	4 MHz	4 MHz	4 MHz
No. of RF channels	25/4 = 6	25/4 = 6	6
Channels/Coverage Area	6/2 = 3	6/2 = 3	6/2 = 3
Control channels/Coverage Area	1	1	1
Usable channels/Coverage Area	3–1 = 2	3–1 = 2	3–1 = 2
SATOP Service per RF channel	1	4*	20**
SATOP channels/Coverage Area	2 × 1 = 2	4 × 2 = 8	20 × 2 = 40
Capacity vs. FDMA	2/2 = 1	8/2 = 4	40/2 = 20
SRAI Index for FDMA	1/20 = 5%		
SRAI Index for TDMA		4/20 = 20%	
SRAI Index for CDMA			100%

**Depends on the number of TDMA slots.*
***Depends on the number of codes.*

Table 1.
Sample calculation of SRAI index for FDMA, TDMA and CDMA [4].

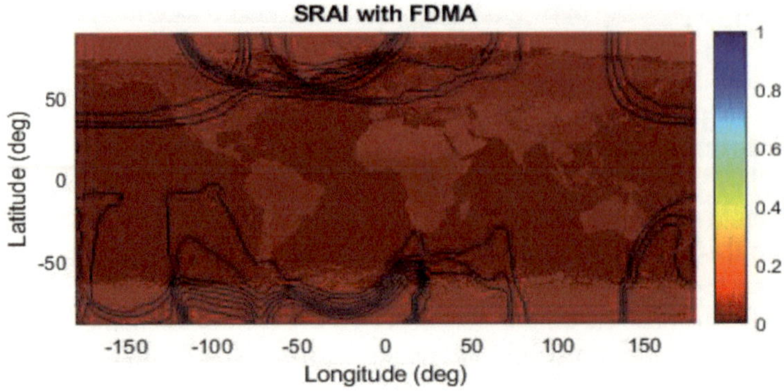

Figure 13.
SRAI simulation results for FDMA.

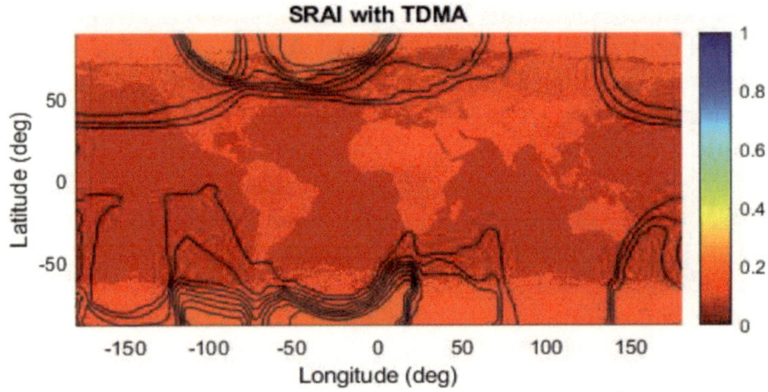

Figure 14.
SRAI simulation results for TDMA.

- **Figure 14**: SRAI for TDMA model described in Section 4.2. For the same notional SOS scenario described in Section 7, the simulation results show that the TDMA allowed more spectrum access than FDMA. The SRAI index is ranging from 0.15 to 0.3.

- **Figure 15**: SRAI model shown in Section 4.2 is modified for CDMA model. Using the same notional SOS scenario described in Section 7, the simulation

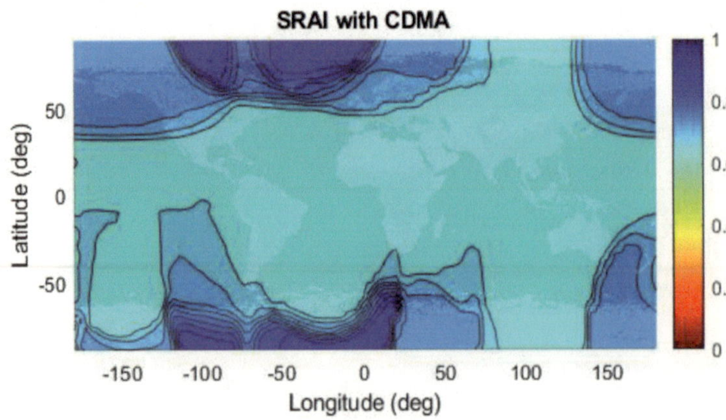

Figure 15.
SRAI simulation results for CDMA.

results show that the CDMA allowed the highest level of spectrum access by allowing more user to share a frequency at once. The SRAI index is ranging from 0.65 to 1.

9. RC implementation in Matlab and simulation results

The RC model presented in Section4.3 was implemented in Matlab, which takes a traditional approach defined by U.D. DOD for modeling avoidance, robustness, recovery, and reconstitution. This model can be broken down into many of the links in a network and they are in any of the given states shown in **Figure 16**. The six possible states shown in **Figure 16** are:

- Radio Frequency (RF) Link achieves maximum capacity as planned
- RF Link avoids the RFI threats with acceptable signal degradation
- RF Link is lost due to increased RFI power
- RF Link has recovered from un-acceptable signal degradation caused by RFI
- RF link has re-established to full-operational capacity while using the 3rd party satellite support
- RF link is disconnected by the users.

The notional SOSE scenario presented in Section 6 was modified for demonstration of the RC model with the following assumptions [8]:

- Simulation time duration: Jan 1, 2020 00:00:00 – Jan 3, 2020 00:00:00
    - 10 second intervals
- SNR($dist$) = $-8\tan^{-1}((dist - 35{,}000)/35{,}000$ “) + 14”
- RFI Power = SNR($dist$) × (1–2φ/100 “)”

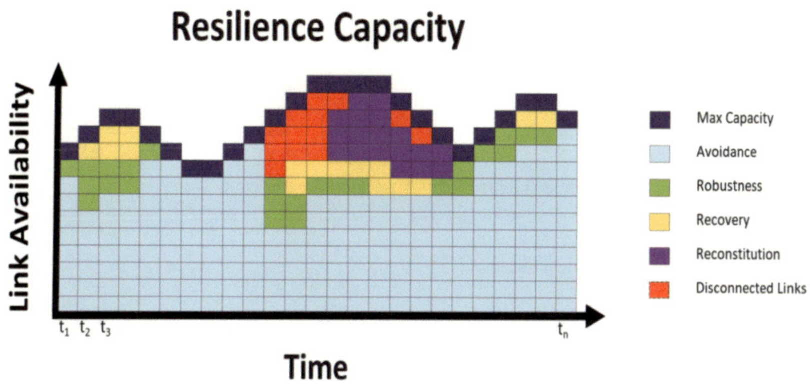

Figure 16.
Potential RC states at any given time.

- For commercial/civilian groups, RFI was not considered as a factor in their internal communication
    - Only the SNR function with Link Margin (LM) > 3
- Only considered RFI on satellites, not ground stations
- If a satellite is only connected to one ground station, then it is not connected
- Ground stations can receive/transmit in different bands, but satellites receive/transmit in same band.

Simulation results show that avoidance was at 74%, robustness at 50% and recovery at 17%. In our simulation, we were able to have full reconstitution of 100%. **Figure 17** shows Network Score[1] for each time step, where each unit of the x-axis represents a 10 second interval from our start time of the 1 sr of January 2020 at time 00:00. The y-axis represents Network Score at each time interval. Whenever a point on the y-axis drops below 1, it means that some portion of ground stations have been cut off from the other ground stations. We observed seven of these downtime windows during our two-day simulation, where RFI from civilian and commercial sources caused a disruption in communication between ground stations via satellite [8].

Figure 18 illustrates seven downtime windows. They are represented on the x-axis. The color bar represents "mean" Network Score over a downtime window.

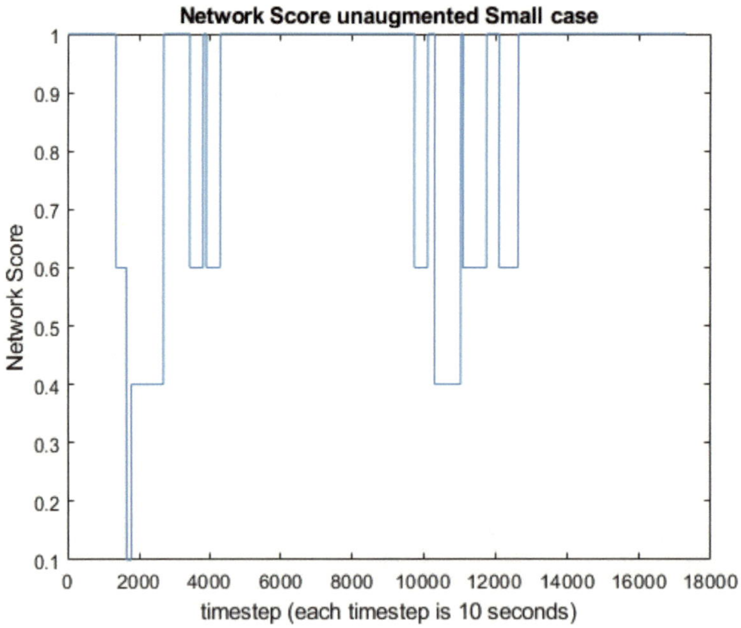

Figure 17.
RC simulation results: all downtime windows.

[1] The network score is defined as the ratio between the number of actual connected network nodes to the ideal number of possible network nodes can be connected. Network score is ranging from 0 to 1, where 1 represents 100% connections in a network.

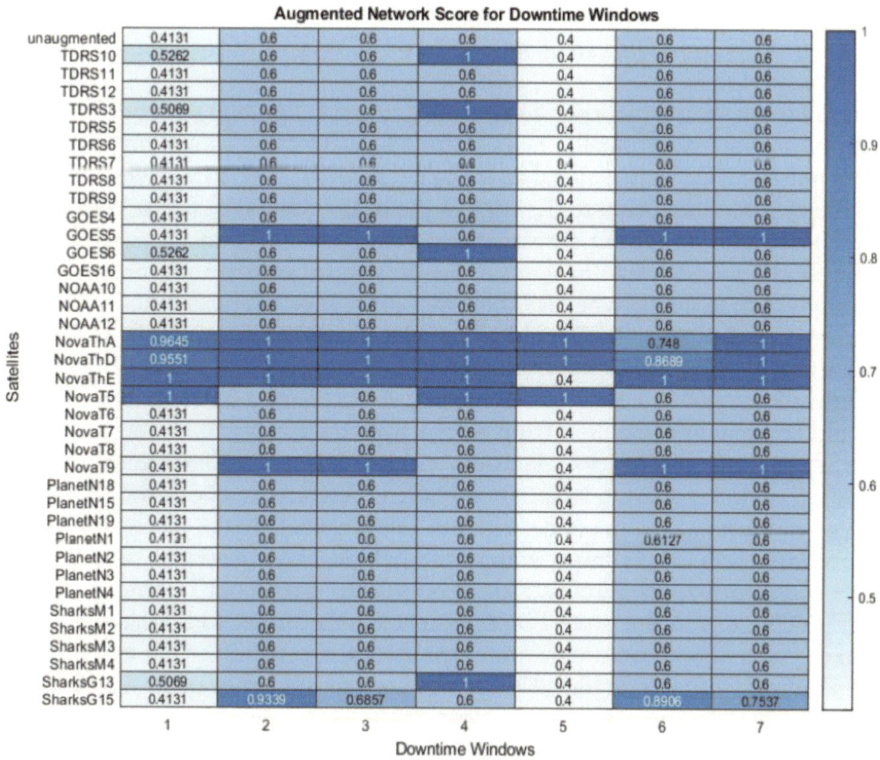

Figure 18.
RC simulation results: seven downtime windows.

The y-axis represents augmentation of the notional military SATCOM network with the labeled satellite during a specific downtime window. The top-most row, labeled "Un-augmented" shows the network status with no additional satellites. The RC model generates these values via the simulation and then searches each column to find the entry with the highest value. It then records this value and which satellite was most supportive to the network during the window. For instance, we find the "mean" network scale for downtime 1 in the first column on the left. The top entry, 0.4131, is the mean network score with no recruited civilian or commercial assisting satellites. We then search down the column noting the maximum of 1 which occurs when the military network is augmented with our theoretical commercial satellite NovaThE. A score of 1 represents full reconstitution of SATCOM network during the window.

10. Conclusion

The RAI-RFI, SRAI and RC models discussed are useful tools in analyzing a space-based SOSE against RFI threats. The models separately address different facets of a space based SOSE. The RAI-RFI model provides a statistical approach for evaluating the best operating network node in the presence of RFI events. The SRAI expands on the RAI-RFI model to account for different spectrum accessing technologies, e.g., FDMA, TDMA or CDMA.

The RC model provides probabilistic analysis of a space based SOSE's ability to successfully maintain communication amidst RFI threats. A network scoring metric has been proposed to assess the state of the network when disconnects occur. The network score is then utilized to identify optimal pre-existing support satellites from third party sources that be leveraged by the military during times of disconnect. Aspects of our model can be refined by expanding on various details that have been oversimplified.

Future work should incorporate an actual dynamic link budget model instead of an approximation. Additionally, more thorough calculations should be considered for RFI noise and the actual signal strength degradation that occurs when moving from the center of the beam to the edge of the beam. Similarly, future considerations should address the cross-polarization isolation that occurs between the signal of interest and the friendly RFI signal. Other important aspects to incorporate might be a recovery time model when RFI becomes present or financial cost as part of the optimization assessment when looking for solutions to the degradation or disruptions caused by RFI. Additional work should also include details regarding existing and future Spectrum Access Systems and Spectrum Sharing Systems. For example, we could implement actual Spectrum Access Systems using FDMA and TDMA. For Spectrum Sharing Systems considerations can be made regarding licensed and unlicensed underlay sharing techniques where secondary users can simultaneously transmit with primary users, and the interference caused by the secondary users over primary users must comply with a threshold level per National Telecommunications and Information Administration (NTIA)/ International Telecommunication Union (ITU) interference protection criterion. Furthermore, additional work should consider actual communication capacities and capabilities per communication link by weighting each link. This can be used to further evaluate Spectrum Access Systems and Spectrum Sharing Systems based on allocated bandwidths. New parameters associated with cognitive radio could also be introduced.

Notes/Thanks/Other declarations

The authors would like to thank the following CSUF graduate students:

- Lauren Benson, Jordan Golemo, and Maria Heinze for their contributions to Section 6 and Reference 6;
- Nicole Hemming-Schroeder and Catherine Osbornefor their contributions to Section 4, Section 7 and Reference 7;
- Scott Digiam battistafor his contribution to Section 9 and Reference 8.

Conflict of interest

The preparation of this chapter was not funded by The Aerospace Corporation, and it was done by the author using his own time and resources, thus it does not represent The Aerospace Corporation's view on the results presented in this chapter.

Author details

Thomas O. Freeze[1*], Tien M. Nguyen[2] and Charles H. Lee[3]

1 The first author performed the work presented in this chapter when he was a Graduate Student at California State University in Fullerton (CSUF). Currently he is with The Aerospace Corporation, El Segundo, California

2 The second author serves as Adjunct Research Professor and CSUF and he is also with The Aerospace Corporation

3 The third author serves as Professor at Program Director of industrial Collaboration Program in Applied Mathematics at CSUF

*Address all correspondence to: tom.freeze@aero.org

References

[1] Fact Sheet: Resilience of Space Capabilities. Department of Defense. https://www.defense.gov/Portals/1/features/2011/0111_nsss/docs/DoD%20Fact%20Sheet%20-%20Resilience.pdf., Jan 11, 2011.

[2] McLeod, Gary. Nacouzi, George. Dreyer, Paul. Eisman, Mel. Hura, Myron. Langeland, Krista S. Manheim, David. Torrington, Geoffrey. Enhancing Space Resilience Through Non-Material Means. Santa Monica, CA, RAND Corporation, 2016.

[3] Edlund, Gregory Military Space Resiliency: Definition, Measurement and Application. Northrop Grumman Aerospace Systems Case 13–1828, Sept. 16, 2013.

[4] Tien M. Nguyen, "Systems-of-Systems Enterprise CONOPS Assessment and Spectrum Resiliency Modeling," Lecture Notes, 2017–2018.

[5] Tien M. Nguyen, Book Chapter titled "SOS Enterprise, SOSE CONOPS, SOSE Architecture Design Approach: A Perspective on Space and Airborne Systems," to be published in the Book titled "Systems-of-Systems Engineering, Modeling, Simulation and Analysis," Intech|Open Publisher, September 2020.

[6] Lauren Benson, Jordan Golemo, and Maria Heinze, "System-of-Systems Enterprise (SOSE) Modeling - SOSE Databases, Data Management and Orbital Dynamics Modeling," Math-597 Final Report, 18 May 2018, California State University in Fullerton.

[7] Nicole Hemming-Schroeder and Catherine Osborne, "System-of-Systems Enterprise (SOSE) Resilience Assessment Index (RAI) Modeling - Modeling of RAI Against Radio Frequency Interference (RAI-RFI) for Ground-Based SOSE Applications with Dynamic Communication Link Margin and Availability," Math-597 Final Report, 18 May 2018, California State University in Fullerton.

[8] Thomas Freeze and Scott Digiambattista, "SOSE Spectrum Resiliency Assessment Index (SRAI) and Resilient Capacity (RC) Modeling: Modeling of SRAI and RC Against RFI Threats for Space-Based SOSE Applications with Dynamic Communication Link Margin and Availability," Math-597 Final Report, 18 May 2018, California State University in Fullerton.

Section 4

SOS Modeling, Simulation and Analysis (MS&A) Applications

Chapter 6

Chapter Review on Computer Simulation of Melt Spinning: A System of Systems Perspective

Derseh Yilie Limeneh and Kelem Tiessasie Yilma

Abstract

This chapter discusses an approach for process simulation in the design of melt spinning process for finding optimal design parameters concerning spinneret, quench air unit and other technical parameters for maximum throughput and quality. The property of as-spun fiber is a function of structural parameters at a given condition and orientation of the structural parameter and it is highly governed by stress level at freeze line. Thus, to define structural property and associated relationship, it requires to identify the process to control the variables (or factors) that affect the structural parameter as well as final fiber property. In addition, this chapter also provides a System-of-Systems (SOS) perspective on melt spinning process and its computer modeling along with mathematical equations for estimating spinline stress with a change in process variables. The spinline stress will be used as an input for a computer simulation to have process optimization by changing the necessary variables until it optimized.

Keywords: melt spinning, system-of-systems, poly ethylene terephthalate, simulation, process parameters, process optimization

1. Introduction

Modeling and Simulation (M&S) have become important tools for evaluating and scheming a melt spinning in a comprehensive arrangement of disciplines fluctuating from fiber spinning and engineering to melt spinning [1]. For example, in engineering scheme, modeling the parameters and simulation used to evaluate the effectiveness of a melt spinning process concept, verify whether all the functional design specifications are meet, or suggest modifications for improving the manufacturability of a product [2]. Melt spinning processes are considered as a system-of-systems process based on the thermoplastic polymer processing method as shown in **Figure 1** [3]. For this process a system is required to ensure the polymer is melted above its melting temperature, and the melted polymer is then transported to another system where it is metered for constant mass through a spinneret into a quench air stream blowing across the spinline [4]. The Spinline-and-Free-line" forming can be considered as another system that is used to form spinline, cool and finally solidify at a distance from the spinneret called 'freeze- line' [5]. The solidified polymer filament is winding at a speed significantly greater than the extrusion velocity; this cause the final cross-sectional area is considerably smaller

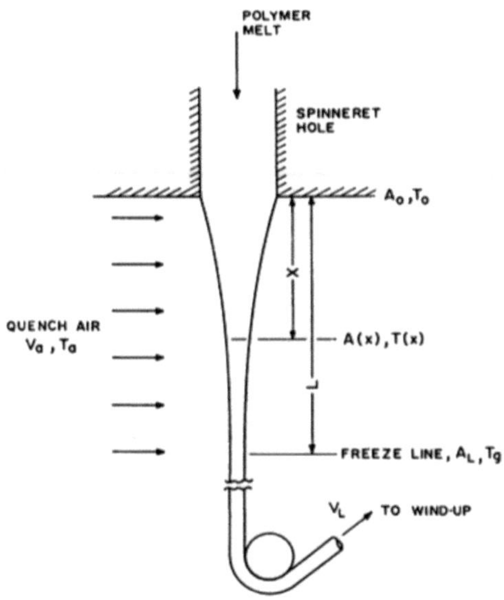

Figure 1.
The melt-spinning process.

(about 100–200 times) than the initial extruded area. This "Winding" process can be considered as another system in the melt spinning process [2, 6]. The way we decompose the melt spinning process into SOS view is based on the "System Control" view-point. For examples: (i) the melted polymer process is considered as a system because it requires a temperature control to ensure the polymer melts above its melting temperature, and (ii) the winding process is considered as another system because it requires to control the winding speed ensuring that the winding speed is significantly greater than the extrusion velocity [7]. Concerning "Spinline-and-Free-line" forming process, note that without considering crystallization in the spinline, solidified filament orientation can be represented by the orientation in the spinning threadline frozen at or near the glass transition temperature of the polymer due to result of tensile deformation under stress in the flowing melt. So, stress level at freeze line can easily govern the orientation, in which the process variables such as extrusion temperature, extrusion velocity, take-up speed, quench air velocity and temperature affect the stress level at freeze line. This shows that molecular orientation in the as-spun filaments is a function of extensional deformation and cooling. Cooling rate across may vary across the filament and cause a change in oriented filament cross section [8]. Thus, "Spinline-and-Freeze-line" system requires a complex system control to ensure accurate orientation, glass transition temperature, and tensile deformation under stress. In practice, the automation in melt spinning can be tested using simulation approach for estimating and gaining insight of the changes of process variables in order to reduce the risk on uniformity of the fiber quality [9].

2. Spinline orientation

Quantification of melt spinning requires the development of mathematical equations describing processing-structural-property correlations. The mathematical equations will be used in the computer simulation for predicting the fiber-line stress using readily measurable process variables such as melt temperature, take-up speed, and extrusion velocity and quantitative correlations between response variable like fiberline stress and structural parameter of the solidified fiber [5]. Using a simple quantitative expressions critical operating conditions and material properties for the melt-spinning for Polyethylene terephthalate (PET) identified. The method used depends on the experimental observation that for vitrified polymers like PET spun at speeds below 3000 m/min, the molecular orientation of the as-spun filament is uniquely determined by stress at freeze line when spinline stress is at the glass transition temperature [9, 10].

Clearly, the spinning stage represents a non-isothermal, uniaxial elongational flow situation with variable physical properties [5]. In order to develop the governing mathematical equations, the following assumptions will be used through this chapter.

1. The process is operating under steady state conditions.

2. The temperature and velocity field are independent of radial position.

3. The elongational viscosity is independent of extension rate and is equal to the Trouton viscosity.

4. All the molecular motions at temperatures less than the glass transition temperature, Tg.

5. The effects of inertia, gravity, surface tension and air drag are negligible.

6. Dieswell has insignificant effect.

Figure 1 shows the coordinate system used for the analysis. Based on the above assumptions, the governing equations can be written as:

$$W = \rho AV \tag{1}$$

$$\sigma = F/A = 3\eta_o dV/dx \tag{2}$$

or alternatively

$$dA/dx = -\rho FA/3W\eta o \tag{3a}$$

and

$$dT/dx = -2(\pi A)^{1/2}/WCp * h(T - Ta) \tag{3b}$$

The appropriate boundary conditions are

$$at\, x = 0, A = Ao, T = To \tag{4a}$$

$$at\ x = L, A = AL, T = Tg \tag{4b}$$

Where L is the distance of the freeze line from the spinneret. The temperature, T (x), the velocity V(x), and the spin line tension F, are obtained by solving Eqs. (2), (3), and (4) subject to the conditions represented by Eqs. (5) and (6). In order to accomplish that, however, correlations for the temperature dependence of physical properties (ρ, Cp 'η_0) and the heat transfer coefficient, h, need to be provided.

Rheological measurements of the shear viscosity of PET indicate that it behaves like a Newtonian liquid for shear rates up to about 200 S^{-1}. Thus, the assumption of

Newtonian behavior may not be too inappropriate for PET [9–12]. The variations of zero shear viscosity of PET, η_0, as a function of intrinsic viscosity (IV) for different temperatures.

$$\eta_0 = 9.76 \times 10^{-3}(\mathrm{IV})^{5.2893} \exp\left[(6923.7)/(\mathrm{T}+273)\right] \tag{5}$$

For density and specific heat, we assume the following linear variations with temperature [13].

$$\rho = 1.375 \times 10^{-3} - 0.75\mathrm{T} \tag{6a}$$

$$\mathrm{Cp} = 9.95 \times 10^{2} + 3.875\mathrm{T} \tag{6b}$$

For the heat transfer coefficient, the best available correlation is [9].

$$\mathrm{h} = 1.98\mathrm{A}^{-1/3}\left[(\mathrm{W}/\rho\mathrm{A})^2 + (8\mathrm{Va})^2\right]^{1/6} \tag{7}$$

A proper choice of F is crucial for smooth operation of the numerical procedure; a fairly good starting estimate can be obtained from the following equation.

$$F_s = 10.635 \frac{\int_{A_0}^{A_L} hA^{-1/2}\mathrm{d}A}{\int_{T_0}^{T_s} \frac{\rho C_p \mathrm{d}T}{\eta_0 (T-T_a)}}, \tag{8}$$

At intermediate take-up speeds (3000 m/min), melt-spinning of PET results in almost amorphous as-spun fibers. That is why molecular orientation of the as-spun fiber is expected to represent the 'frozen-in' stress in the spinline [1, 2, 14]. Molecular orientation which is measured by birefringence quantitatively related to process variable form a link between the latter and the as-spun fiber properties.

Experimental data shown in **Figure 2**, for melt-spun PET fibers under different process conditions demonstrate that the birefringence is directly proportional to the spinline stress for PET,

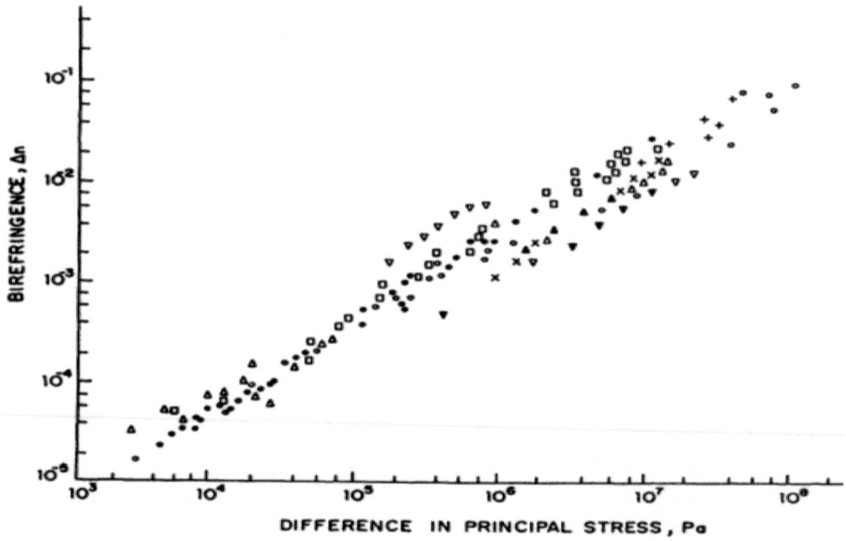

Figure 2.
Variation of birefringence with principal stress difference for PET with + = Mn ~ 24,000; ∇ = Mn on-line [14]; X = Mn 15,000; ○ = Mn, 17,000.

$$\Delta n = C\sigma = C(F/A) \tag{9}$$

C being the stress optical coefficient [9]. The intermediate take-up speeds in a melt-spun PET fibers show amorphous at a slow crystallization behavior of PET, and the birefringence depends on the (experimentally measured) spinline stress. To show the relationship between as-spun fiber birefringence and the calculated stress at the freeze line, the stress at freeze line is taken to be the spinline stress at which the velocity is 95% of the take-up velocity [5, 15]. It is observed that a linear variation is obtained with stress optical coefficient approximately about 6×10^{-9} to 9×10^{-9} Pa $^{-1}$, which is then compared with the values obtained by different experiments [5]. For other experiments, the spinline stress controls the orientation of the as-spun fiber, which in turn determines the mechanical properties and associated fiberline processing ability. Thus, it is clear that stress at freeze-line controls the orientation of the as-spun fiber, and in turns it can determine the mechanical properties [4, 13]. The sensitivity of the as-spun fiber properties is directly related to the sensitivity of freeze-line stress to the changes in process variables. Stress at freeze-line, on the other hand, can readily be obtained from the process simulation calculations, provided the operating conditions and the melt properties are available [16].

3. Sensitivity analysis

Sensitivity analysis provides an approach to look into the effect of a factor affecting the response variables and making effective changes in the process parameters for controlling process stability and product quality improvement [1, 6]. As shown in **Table 1**, the critical process variables have a significant effect on stress level at the freeze-line are quantified for a typical PET spinning process [17]. The spinline stress is calculated under different operating conditions and melt properties, which is then used in the assessment of stress sensitivity, are shown in **Figures 3** and **4** below [9].

The results indicate that the changes in process variables like extrusion temperature, melt, take-up velocity and the melt throughput rate that can affect the response variables (e.g., stress at the freeze-line) [5, 8]. The slope of the curve provides a measure of the sensitivity of stress at the freeze-line to process variable

Process variable	Value
Intrinsic velocity	0.611 dl g^{-1}
Extrusion temperature	285°C
Glass transition temperature	67°C
Ambient air temperature	25°C
Quench air velocity	48 m min^{-1}
Spinneret hole diameter	0.00038 m
Take-up velocity	1500 m min^{-1}
Mass flow rate	0.001 kg min^{-1}
Spinline tension[a]	0.000972 N
Freeze line location from spinneret	0.37 m

[a]*Calculated value.*

Table 1.
Process variables for PET in a typical melt-spinning operation.

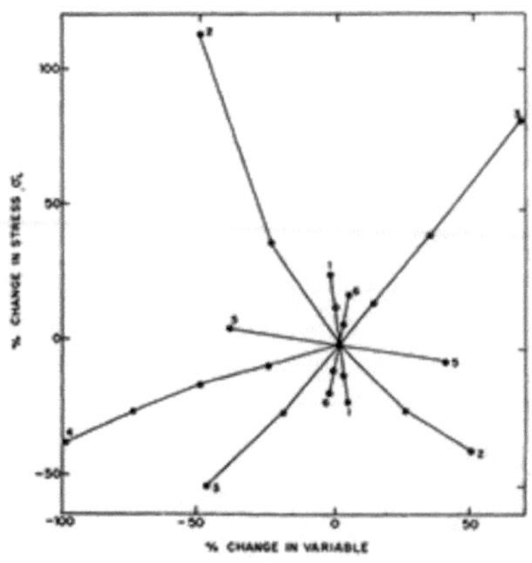

Figure 3.
Sensitivity plot for stress at the freeze-line. The various curves are for percentage changes in (1) extrusion temperature, (2) melt flow rate, (3) take-up velocity, (4) quench air velocity, (5) quench air temperature, and (6) melt intrinsic viscosity [5].

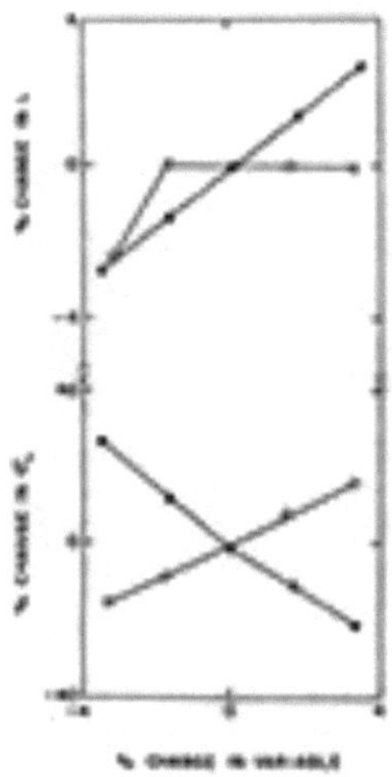

Figure 4.
Expanded sensitivity plot for extrusion temperature = ●, and intrinsic viscosity = ○ [9].

changes. In comparison, stress at the freeze-line is moderately sensitive to changes in quench air velocity and relatively insensitive to alterations in the quench air temperature [9]. But it is more sensitive to a changes in the throughput rate as well as the extrusion temperature and is relatively unperturbed by changes in the melt intrinsic viscosity. But other changes exert only moderate influence in determining the freeze-line location [6]. Since the as-spun fiber quality is uniquely determined by its birefringence, which in turn is directly related to stress at the freeze-line, so the degree of sensitivity of the as-spun fiber properties changes in these process variables are the same as that of stress at the freeze-line [15, 18, 19].

4. Process implications

The information obtained from the sensitivity analysis provides good insight for identifying the critical process with associated level of risk [14]. The process for controlling the as-spun filament orientation requires critical process parameters like extrusion temperature, intrinsic viscosity of the polymer melt take-up velocity and melt flow rate. These parameters need to be taken into consideration, unless the control process quality is having issues as discussed below [5].

If there is an increase in temperature and/or a decrease in melt intrinsic viscosity, the spinline stress is being easily reduced, which affects the stretch ability as well as final fiber quality [20]. These higher sensitivity of the spinline stress in response to extrusion temperature has important implications for maintaining uniform product quality in a multiposition commercial plant with each position carrying multiple hole spinnerets [21, 22]. As shown in **Figure 5**, Curve 4, a variation of about 3°C in quench air temperature profile of the spinneret represents a 1% change in the extrusion temperature; but this could result in about 10% variation in the spinline stress by increasing the coefficient of variation of elongation [9]. So, having static mixers upstream of the spin pack is the best way for delivering a thermally homogeneous melt to the spinnerets [21–23]. Both the melt flow rate and the take-up velocity can also affect as-spun filament quality. This is due to, increase in take-up velocity at a constant flow rate results in increase orientation due to enhanced stretching. But for a constant take-up velocity, the orientation is reduced as the flow rate is enhanced. Because the filament denier increases rapidly with increase in flow rate, and even the spinline tension actually decreases proportionally with a net effect in a reduction in spinline stress as throughput is increased [14, 20]. By controlling such process variables (i.e., the melt flow rate and the take-up speed), it is possible to produce products over a range of denier and tenacity if needed [9, 17]. In addition, if the process is a batch process, the moisture content needs to be controlled less than 0.005% and even a moisture level of 0.01% could lower the polymer intrinsic viscosity from 0.6 to 0.56. In case if the change in intrinsic viscosity is 7%, which leads to a 35% change in spin line stress, this thereby affecting spinnability and product quality that desired [5].

In plants with continuous polymerization, variation in polymer intrinsic viscosity may also result from reactor perturbation [22]. Therefore, incorporation of

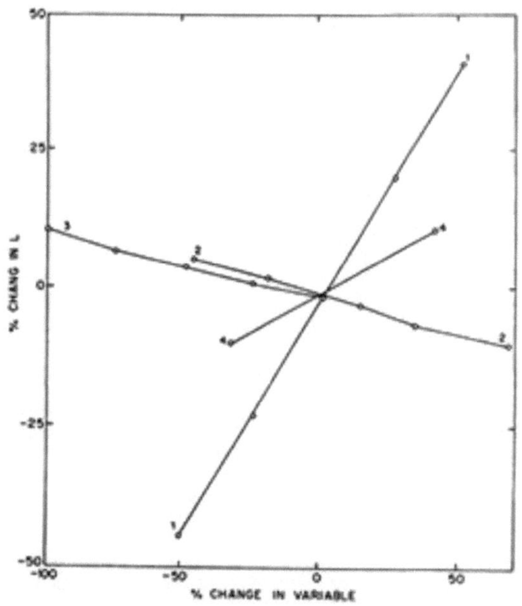

Figure 5.
Sensitivity plot for freeze-line location. The various curves are for changes in (1) melt flow rate, (2) take-up velocity, (3) quench air velocity, and (4) quench air temperature [9, 14].

an on-line viscometer to monitor the melt quality is likely to be a necessary instrumentation. There are different ways of increasing productivity. One approach is to use a process optimization discussed above, i.e., by changing the process variables to find optimum solution. The other approach is to increase in the number of holes in the spinneret by 30% [24]. If possible, as per hole layout guidelines depend on the investment to manipulate such multi-hole spinneret [9, 23, 24]. The location of the freeze-line has critical practical implications for spinline stability and product uniformity in multifilament spinning air turbulence [1, 9, 22]. An incorrect location of the freeze-line can lead to filament breakage and/or fused filaments resulting in process interruptions and poor product quality. Reducing the quench air temperature from 25 to 15°C show 40% change in the filaments stability by moving the freeze-line closer to the spinneret without a significant effect on the structural parameter like orientation, this makes it be fewer disturbances to air [9]. The take-up velocity is not the most dominant controlling variable for orientation development since its process implication does not have much impact on spun-fiber.

5. Stress-orientation relationship for PET

The Dutta Nadkarni simulation package is recommended for simulation of the melt spinning process [9, 20, 23]. Before we use the computer simulation for spinning process optimization, it is important to develop a way to quantify the correlation between the spinline stress and the as-spun filament orientation, such that the operator can confirm its validity over a wide range of spinning process parameters in current industrial operations is the data are collected from three different plants for computing the spinline stress, stress at the freeze-line [5, 25]. **Table 2** summarizes the industrial data collected. Using the data birefringence's, of the as-spun filaments orientation are measured experimentally (**Figure 6**).

$$\sigma_L = 2.526 \times 10^{-9} \left(\Delta n^{1.265} \right) \tag{10}$$

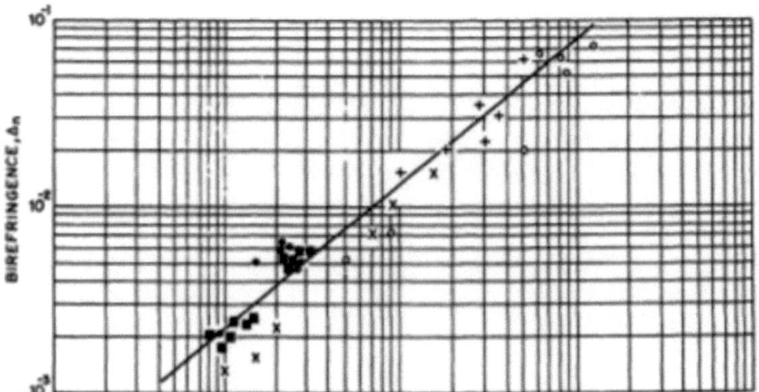

Figure 6.
Variation of birefringence with principal normal stress difference at the freeze-line. Filled symbols represent data collected from various industrial operations from India while the other symbols represent other published data [9]. The birefringence's value is correlated with linear regression analysis to fit the data and the correlation is obtained with a correlation coefficient of 0.96 for fitting 37 data points as shown in ***Figure 6*** *(straight line fit between birefringence and stress at the freeze line on the log–log plot) and Eq. (13) [5].*

Process parameter	Range	No. of variables
Intrinsic velocity	0.6–0.65	2
Extrusion temperature	275–290	3
Quenching air temperature	15–25	7
Quench air velocity	30–130	3
Spinneret hole diameter	0.2–0.3	11
Take-up velocity	500–1250	9
Throughput rate	0.5–1.0	2
Filament diameter	4.0–12.0	11

Table 2.
Parametric space covered (taken from actual industrial data).

The above equation is used as an input to the melt spinning process for optimizing the relationship between birefringence's and spinline stress for the melt-spinning of PET at speeds up to 3000 m min^{-1} with a correlation coefficient of 0.96.

6. Case study for process optimization

The use of Eq. (1) along with the recommended computer simulation package for a case study to: (i) Increase the production rate, (ii) Quality of the fiber, and (iii) a new desired product developed through process optimization in spinning without affecting the fiber line processing conditions [9, 23]. Maintain the denier and orientation of the as-spun filaments at the reference levels because for a given denier product, the throughput rate and the take-up speed have to be increased proportionally which avoid downstream process optimization at the higher productivity [15]. For cases to increase in productivity changing the polymer, viscosity is not desirable rather other process parameters are desirable, such as extrusion temperature, quench air velocity and quench air temperature [7, 26]. According to the process sensitivity analysis, the orientation of the as-spun filaments, as measured by birefringence is very sensitive to changes in the spinning temperature and since the orientation related to freeze-line stress, the spinning temperature is to be changed so as to maintain freeze line stress and spin line tension at the original values [21, 25]. Quench air velocity and temperature are the other process variables that can be changed but they do not show significantly affect the spinline stress, rather on the freeze-line location in order to ensure stability of the spinline to avoid denier variation and final fiber quality [3, 20, 27]. The production rate can also be improved by a proportional increase in the number of spinneret holes with no changes in throughput per hole, take-up speed and spinning temperature and even it is best option than process optimization through changing variable [11].

The reference process parameters of a spin- line whose productivity is to be improved are summarized in **Table 3** [9] along with the process conditions determined for production rate increases of 10, 20 and 30%. The process variables provided in this table are representative of values used in a typical spinning operation, as averaged out from within the ranges given in **Table 2**. The computer simulation of Dutta and Nadkarni is used for determining the changes in the process variables when the throughput rate and take-up speed changes by 10, 20 and 30% and the effect on the response parameters, including spinline stress, spinline tension and freeze-line location, for increased productivity is shown in **Figure 7** [11].

Parameter	Original condition	Increase in melt flow rate, take up velocity(W,VL)		
		10%	20%	30%
Intrinsic velocity	0.6	0.6	0.6	0.6
Extrusion temperature	280	280	280	280
Quenching air temperature	20	20	20	20
Quench air velocity	100	100	100	100
Spinneret hole diameter	0.2	0.2	0.2	0.2
Take-up velocity	1000	1100	1200	1300
Throughput rate	0.5	0.55	0.6	0.65
Spinline stress	2.18	2.32	2.46	2.60
Spin line tension	82.8	87.5	93.8	98.0
Freeze line location	26.4	28.0	30.8	33.0

Table 3.
Effect of changes in productivity on spinline stress, tension and location of the freeze line [14].

All process variables show proportional increments with increased throughput and take-up velocity [15]. Thus, to maintain the molecular orientation and the downstream process variables constant, it is essential to have the values of spinline stress, spinline tension and freeze-line location with proper adjustment of other process variables [25]. The sensitivity analysis indicates that increasing extrusion temperature decreases the spinline stress and spinline tension but extends the freeze-line location as shown in **Figure 8** below.

When extrusion temperature is assumed to be equal to the spinneret exit temperature, there is a 10% increase in productivity. In order to return the spinline stress and tension from a 6% increase to the original level, the increase in the extrusion temperature should be approximately 1.87% [5, 7]. However, a 1.87% increases in extrusion temperature would result in a further 1.75% increase in Freeze-line location. The freeze-line location is thus sensitive to the combined

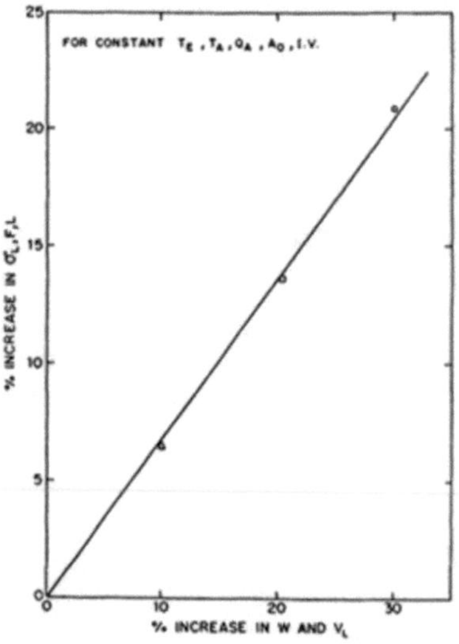

Figure 7.
Simulation results showing effect of percentage increases in throughput and take-up velocity on the spin-line stress, spin-line tension and freeze-line location.

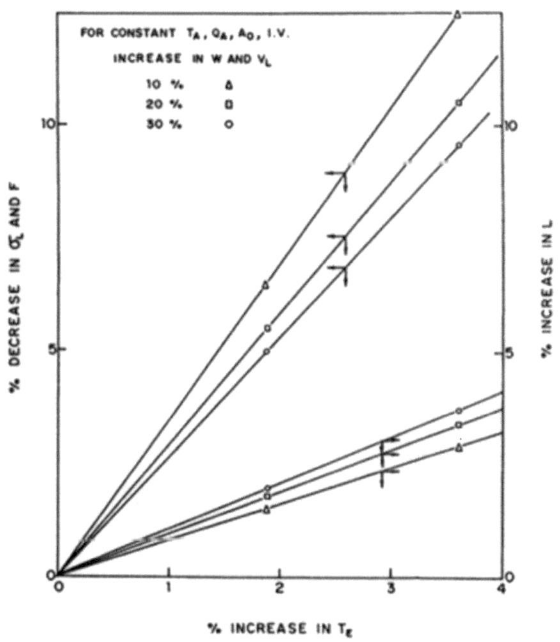

Figure 8.
Effect of percentage increase of extrusion temperature on the spinline stress, spinline tension and freeze-line location for each of 10%, 20% and 30% increase in throughput and take-up velocity.

changes in throughput rate and extrusion temperature. [17, 24] Referring to **Table 4**, although the spinline stress and tension can be brought back to the original values by a change in the extrusion temperature, the freeze-line location remains

Parameter	Reference condition intrinsic viscosity (IV) = 0.6	10% decrease in freeze line location, take up velocity (L,W)	1.9% increase in extrusion temperature (T_{E})	45% decrease in quench air temperature (Ta)
Intrinsic velocity	0.6	0.6	0.6	0.6
Extrusion temperature	280[a]	280	285	285[a]
Quenching air temperature	20[a]	20	20	11[a]
Quench air velocity	100	100	100	100
Spinneret hole diameter	0.2	0.2	0.2	0.2
Take-up velocity	1000[a]	1100	1100	1100[a]
Throughput rate	0.5[a]	0.55	0.55	0.55[a]
Spin line stress	2.18[b]	2.32	2.20	2.19[b]
Spin line tension	82.8[b]	87.9	83.3	82.9[b]
Freeze line location	26.4[b]	28.0	28.9	26.8[b]

[a]*Parametric values changed to achieve a 10% increase in productivity for constant product quality.*
[b]*Values held constant to maintain product quality.*

Table 4.
Summary of process changes required obtaining a 10% productivity increase while product quality is maintained constant.

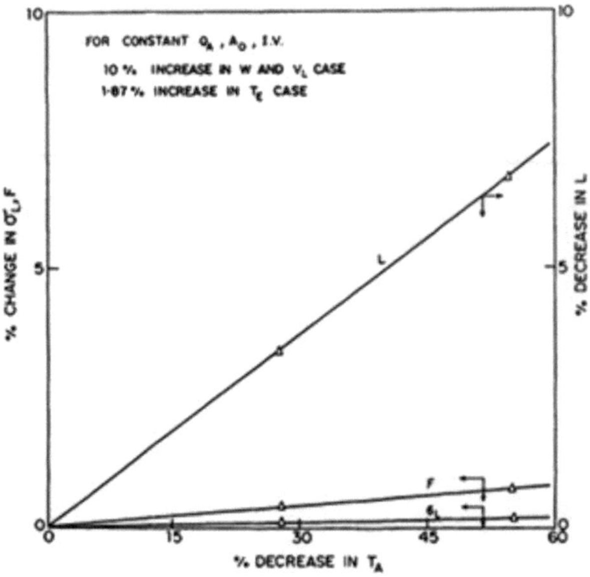

Figure 9.
Effect of percentage decrease in quench air temperature on spinline stress, spinline tension and freeze line location for the specific case of 10% increase in throughput and take-up velocity [9, 24].

different from the original value by about 9.5%. The location of the freeze-line can affect the spinline stability and product uniformity [5, 25]. Hence its retention at the original value is desirable. A process variable is to be sought, which would move the location of the freeze line closer to the spinneret without any significant changes in either spinline stress or Freeze-line location. The quench air temperature is known to influence freeze-line location to a greater extent relative to spinline tension and spinline stress and can be effectively used for stabilizing the spinline. From **Figure 9** it is seen that to move the freeze- line location by about 8.0%, the quench air temperature has to be decreased by about 45%. Furthermore, the effect of this process change on spinline stress and freeze line location is insignificant. The entire exercise is summarized in **Table 4**, which indicates the changes in process variables required to increase productivity while maintaining the denier, orientation and stability of the spinline [5].

7. Case study for product development of low pill fibers

Lowering the intrinsic viscosity or molecular weight by having drawn fiber denier of 1.2 held constant to make the downstream process variables at constant values [12, 28] (**Table 5**). Because of this the changeable variables, it is required to lower the draw ratio to 3.5 from reference value to prevent fiber breakage in downstream process and this change as-spun denier from 4.5 to 4.2. Due to 8.3% decreases in intrinsic viscosity a new pill fiber is developed. However, this phenomena (decrease in spun fiber diner) causes the decrease in Spinline stress and Spinline tension, which makes melt unspinnable increases the take-up velocity and the throughput rate to new values (22.5%, 14.2% respectively), and lowering extrusion temperature to 275°C. This is a way to increase the Spinline tress with as-spun denier at a value of 4.2. Freeze-line location has effect on the spinline stability and product uniformity stress [29]. However having 50% decreases in quench air temperature from reference value, to maintain the spinline stress and freeze-line

Parameter	Reference condition intrinsic viscosity (IV) = 0.6	8.3% decrease in melt intrinsic viscosity (IV)	22.5% increase in take up velocity (V_L) 14.2% increase in melt flow rate (W)	1.8% decrease in extrusion temperature (T_E)	50% decrease in quench air temperature (T_a)
Intrinsic velocity	0.6[a]	0.55	0.55	0.55	0.55[a]
Extrusion temperature	280[a]	280	280	275	275[a]
Quenching air temperature	20[a]	20	20	20	10[a]
Quench air velocity	100	100	100	100	100
Spinneret hole diameter	0.2	0.2	0.2	0.2	0.2
Take-up velocity	1000[a]	1000	1225	1225	1225[a]
Throughput rate	0.5[a]	0.5	0.571	0.571	0.571[a]
Spin line stress	2.18[b]	1.63	2.05	2.16	2.18[b]
Spin line tension	82.8[a]	61.8	72.5	76.6	76.9[a]
Freeze line location	26.4[a]	26.3	29.6	29.3	26.5[a]
As-spun fiber denier	4.5[a]	4.5	4.2	4.2	4.2[a]
Drawn fiber denier	1.2[b]	1.2	1.2	1.2	1.2[b]
Draw ratio	3.75[a]	3.75	3.5	3.5	3.5[a]

[a]*Parametric values changed to achieve low pill fiber product.*
[b]*Values held constant to maintain the downstream process variables at constant values.*

Table 5.
Summary of process changes required obtaining low pill fibers without affecting downstream variables [9].

location at the original values, we need to move the location of the freeze-line closer to the spinneret [9, 11, 12, 30].

8. Conclusion

The mechanical property of fiber is highly governed by the orientation level in the thread line. However, this structural parameter is heavily affected by the melting process as shown in the sensitivity analysis above with different level of effects. According to the above analysis, it is shown that the molecular orientation developed in a melt flow in spinning process is governed by the spin-line stress (stress at freeze line). A mathematical equation, which is an input for the simulation is developed according to a given quantitative data, shows a linear relationship between the orientation and stress levels. The simulation results are correlated with the variables associated with the spin-line stress to determine if the stated assumptions are in agreement with the simulation results. The combination of the process simulation and stress-orientation relationship gives us a procedure for identifying and simulating the important spinning process variables that would affect the spin-line stress, and hence the mechanical properties and fiber line processability of the as-spun filaments. Once the effect is simulated the process optimization is done by changing the process variables according to the final property desired. Therefore, the process optimization can be achieved using SOS perspective and computer simulation. This is a good advancement in melt spinning to achieve the process optimization. The future work will tell on computer simulation and modeling of manmade fibers and their parameters based on systems of system perspective.

Notations

A_O	initial cross sectional area
T_O	spinneret exit temperature
A(x)	cross sectional area at (x) distance
T(x)	spinneret temperature at (x) distance
Va	quench air velocity
V_L	final velocity
Ta	quenches air temperature
A_L	final cross sectional area
X	distance from spinneret
L	freeze line location
Tg	glass transition temperature
n	birefringence
σ_L	stress at freeze line

Author details

Derseh Yilie Limeneh* and Kelem Tiessasie Yilma
Ethiopian Institute of Textile and Fashion Technology, Bahir Dar University, Bahir Dar, Ethiopia

*Address all correspondence to: derseh2003@gmail.com

References

[1] Baran I, Cinar K, Ersoy N, Akkerman R, Hattel JH. A review on the mechanical modeling of composite manufacturing processes. Archives of Computational Methods in Engineering. 2017;**24**(2):365-395

[2] Barshop BA, Wrenn RF, Frieden C. Analysis of numerical methods for computer simulation of kinetic processes: Development of KINSIM—A flexible, portable system. Analytical Biochemistry. 1983;**130**(1):134-145

[3] Bungartz HJ, Schafer M. Modelling. Simulation. Optimization. 2014;**53**

[4] Doufas AK, McHugh AJ, Miller C. Simulation of melt spinning including flow-induced crystallization: Part I. Model development and predictions. Journal of Non-Newtonian Fluid Mechanics. 2000;**92**(1):27-66

[5] Gupta VB, Kothari VK e. Manufactured fiber technology. In: Springer Science & Business Media. 2012

[6] Mattrand C, Beakou A, Charlet K. Numerical modeling of the flax fiber morphology variability. Composites Part A: Applied Science and Manufacturing. 2014;**63**:10-20

[7] Spagele T, Kistner A, Gollhofer A. Modelling, simulation and optimisation of a human vertical jump. Journal of Biomechanics. 1999;**32**(5):521-530

[8] Lin S, Ryu S, Tokareva O, Gronau G, Jacobsen MM, Huang W, et al. Predictive modelling-based design and experiments for synthesis and spinning of bio inspired silk fibres. Nature Communications. 2015;**6**(1):1-12

[9] Nadkarni, V.M., 1997. Computer simulation of melt-spinning. In Manufactured Fibre Technology (pp. 98–123). Springer, Dordrecht

[10] Singha K, Maity S, Singha M, PAL S. Computer simulations of textile non-woven structures. Frontiers in Science. 2012;**2**(2):11-17

[11] Schutze M, Butler D, Beck BM. Modelling, Simulation and Control of Urban Wastewater Systems. Springer Science & Business Media; 2011

[12] Shengwang Y. Influence of TCS process conditions on pet FDY formation [j]. China Synthetic Fiber Industry. 2001;**5**

[13] Cao C, Chung TS, Chen SB, Dong Z. The study of elongation and shear rates in spinning process and its effect on gas separation performance of poly (ether sulfone) (PES) hollow fiber membranes. Chemical Engineering Science. 2004; **59**(5):1053-1062

[14] Bell WP, Edie DD. Measured orientation and internal stress distributions in melt-spun fibers. Journal of Applied Polymer Science. 1987;**33**(4):1089-1102

[15] Kikutani, T., 2009. Structure development in synthetic fiber production. In Handbook of Textile Fibre Structure (pp. 157–180). Woodhead Publishing.

[16] Muller-Plathe F. Molecular dynamics simulation of gas transport in amorphous polypropylene. The Journal of Chemical Physics. 1992;**96**(4): 3200-3205

[17] Hehenberger P, Vogel-Heuser B, Bradley D, Eynard B, Tomiyama T, Achiche S. Design, modelling, simulation and integration of cyber physical systems: Methods and applications. Computers in Industry. 2016;**82**:273-289

[18] Clancy TC, Mattice WL. Computer simulation of polyolefin interfaces.

Computational and Theoretical Polymer Science. 1999;**9**(3–4):261-270

[19] Marla VT, Shambaugh RL. Three-dimensional model of the melt-blowing process. Industrial & Engineering Chemistry Research. 2003;**42**(26): 6993-7005

[20] Bhuvanesh, Y.C. and Gupta, V.B., 1990. Computer Simulation of Melt Spinning of Poly (Ethylene Terephthalate) Using a Steady-State Model.

[21] Cicek, H., 1990. Computer Simulation of Melt Spinning.

[22] Elser M, Tschegg EK, Finger N, Stanzl-Tschegg SE. Fracture behaviour of polypropylene-fibre-reinforced concrete: Modelling and computer simulation. Composites Science and Technology. 1996;**56**(8):947-956

[23] Toth R, Coslanich A, Ferrone M, Fermeglia M, Pricl S, Miertus S, et al. Computer simulation of polypropylene/organoclay nanocomposites: Characterization of atomic scale structure and prediction of binding energy. Polymer. 2004;**45**(23): 8075-8083

[24] Huang N, Yu X, Tang Z, Hou P, Xu X. Computer simulation of polyester high-speed thermal channel spinning. Macromolecular Materials and Engineering. 2002;**287**(5):330-338

[25] Gagon DK, Denn MM. Computer simulation of steady polymer melts spinning. Polymer Engineering & Science. 1981;**21**(13):844-853

[26] Liang X, Ding Y, Wang Z, Hao K, Hone K, Wang H. Bidirectional optimization of the melting spinning process. IEEE transactions on cybernetics. 2013;**44**(2):240-251

[27] Bhuvanesh YC, Gupta VB. Computer simulation of melt spinning of polypropylene fibers using a steady-state model. Journal of Applied Polymer Science. 1995;**58**(3):663-674

[28] Cooke WD. 21—Fibre fatigue and the pilling cycle part iii: Pill wear-off and fabric attrition. Journal of the Textile Institute. 1984;**75**(3):201-211

[29] Ukponmwan JO, Mukhopadhyay A, Chatterjee KN. Pilling. Textile Progress. 1998;**28**(3):1-57

[30] Sriprateep K, Bohez ELJ. A new computer geometric modelling approach of yarn structures for the conventional ring spinning process. The journal of the textile Institute. 2009; **100**(3):223-236

Chapter 7

Simulating the Effect of Friction on Drive Screw Using System-of-System Modeling with Predetermined Torque

Fatima Isiaka, Awwal M. Adamu, Salihu A. Abdulkarim and Abdullahi Salihu

Abstract

In most mechanical systems, screw threads serve three main basic purposes: (i) to transmit power, (ii) to provide a clamping force, and finally (iii) to restrict or control motion. This chapter demonstrates the effects of friction and behavior which can occur in a bolted fastening (screw thread) for advanced design purposes. To model this behavior, other control components are attached to the bolted screw. The bolt preload is applied with a predetermined torque. For this case the preload depends on the friction under the head and in the thread. The friction prevents the loosing of the bolted fastening. This effect is termed as self-locking effect. We designed an algorithm that reproduces an exemplary simulation scenario, which determines friction and its effect on thread angle based on the strength of the coefficient of friction at a specific tension or clamp load value using the system-of-system approach. The result shows specific behavior on both the motion in threads and drive screw with predetermined torque. The chapter is limited to creating a simple simulation environment to demonstrate the effects.

Keywords: predetermined torque, bolted fastener, tension, runout turn, drive screw

1. Introduction

A drive screw usually consists of head radius, shank, runout, and threads (as shown in **Figure 1**). The chapter defines the behavior of a thread fastener on drive screw by focusing more on the thread behavior and on the fasteners [1]. The mating parts introduced in this chapter consist of the worm drive, which meshes with the worm gear connected to the drive screw and the hypoid, to maintain high-speed reduction using a pair of hypoid gears. They are used to control friction by reducing rotational speed and transmit higher torque. Usually, when a bolt is tightened by applying a torque to the bolt head, the bolt gets stretched out causing the clamping plates to be compressed together. Fastener coatings are being required to comply with corrosion specifications and also fulfill specified tightening characteristical requirements. Some of these requirements mostly include test for torque tension and sometimes a coefficient of friction test. The torque-tension test usually consists of an acceptance input

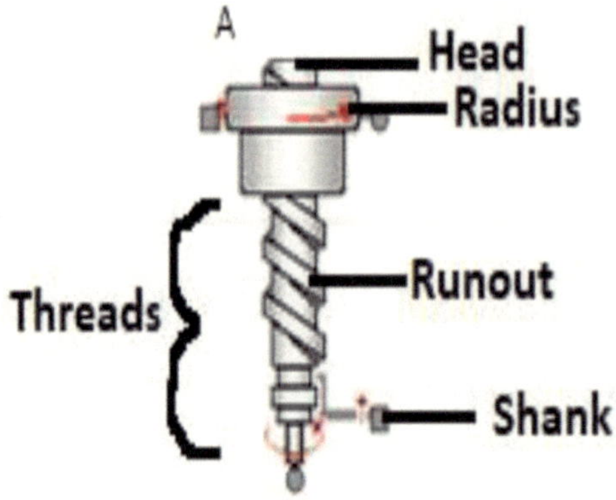

Figure 1.
A typical drive screw with its characteristic component.

torque which is often known as the predetermined torque at a specific tension clamp load value, while the coefficient of friction tests requires an acceptance coefficient of reference friction with a specified tension or a clamp value [2, 3].

Threaded faster assembles take some process, most especially for critical bolted joints, which takes some stages into account; this involves controlling both the input torque and the angle of tension that turns to achieve the desired output of proper preload of the bolted assembly [4, 5]. An accurate clear understanding of the role of friction is needed for both the underhead and threaded contact zones to giving a definition to the relationship between the angle of tension, the predetermined torque, and tension control; there are a lot of factors that affect the tension created in a bolt when tightening behavior occurs in the torque control. All the bolted joints consist of the main torque-angle signature. Additionally, by combining the torque-angle curves with a precise computation and a basic understanding of the logic mechanisms of the threaded fasteners, we can obtain the practical information required to evaluate the characteristics of the individual fastener tightening processes. The torque-angle curves can provide the necessary parametric data to properly qualify the capability of tightening tools to tighten a given fastener [6–8].

Figure 2 shows the energy transfer process that occurs when tightening threaded fasteners. The area under the torque-angle curve is basically proportional to the energy required to tighten the fastener. To model this process, achieving proper control of the tightening process is possible if we get a comprehensive understanding of the relationship between the torque and turn in the development of the tension [9, 10]. It is necessary to be familiar with what happens when a typical fastener is tightened. The concept behind tightening a fastener involves turning in advance the lead screw and torque at turning moment so that the preload tension happens at the fastener. The predicted output is a clamping force that holds the components together. Modeling the torque turn signature for the fastener tightening process has four characteristic zones (**Figure 3**). The first is the rundown or prevailing zone, which occurs before the fastener head or nut contacts with the bearing surface. Basically, the alignment zone is the second part of the tightener; in this zone, the fastener and the joint mating surfaces are drawn into alignment to obtain a possible snug condition. The elastic clamping is the third zone, where the slope of the torque-angle curve is alternatively constant [11, 12]. The last zone, which is the post-yield zone, begins with an inflection point at the end of the elastic range. Constantly, this fourth zone can be due to yielding in the joint or gasket or due to the yield of the threads in the nut or clamped components or nut rather than

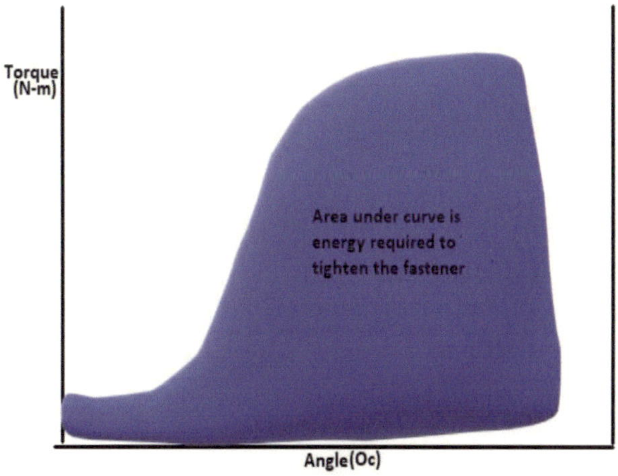

Figure 2.
Total energy area required to tighten the bolt fastener.

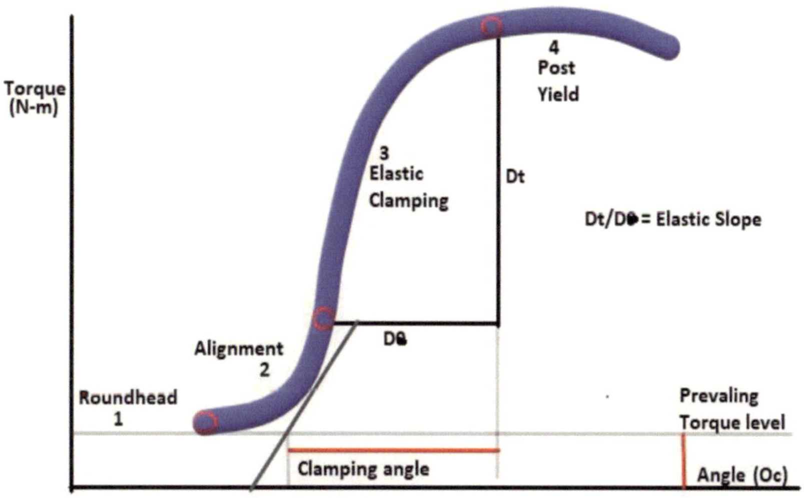

Figure 3.
A basic narrative of the four zones in a threaded screw.

to the yield of the fastener. This chapter is mostly focused on what occurs with the threads, when all zones are combined in the tightening process [13, 14].

2. The system-of-system approach

Drive screws is mostly a subject area less discussed in mechanical system methodology. Most of its process optimization, especially on fastener tightening, is purely theoretical and not deployed dynamically, which is mostly controversial because of its limited area of application. The application of system of systems (SOS) is a new approach to addressing this particular area, by connecting its basic components to different controls to optimize its process flow. The controls or process blocks are obtained from the visual library of simulation X, which enables the set of tools for deployment and development processes. This approach or method requires various blocks, which seeks to optimize each individual controls or

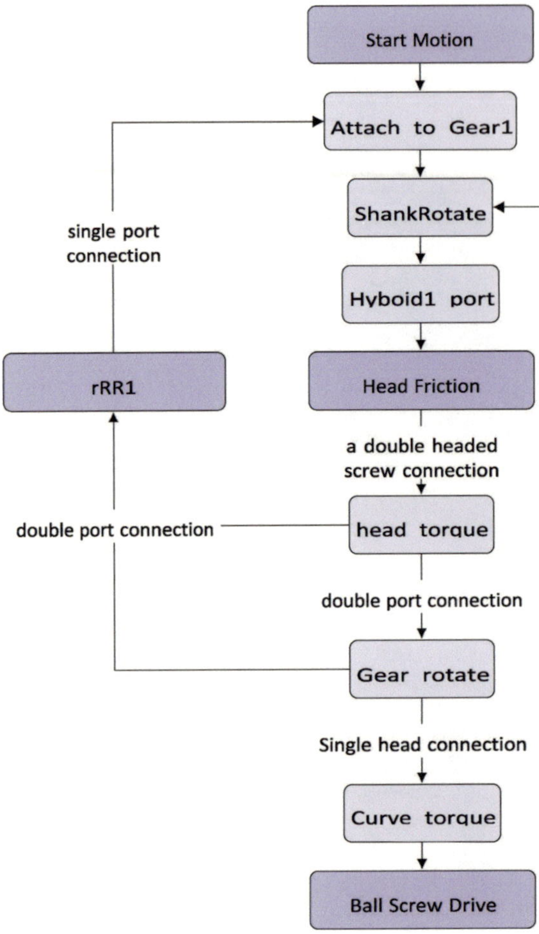

Figure 4.
Algorithm to determine the logic flow for a drive screw.

process flow. The tension of fasteners and torque effect are two different aspects of this system which are discussed in Sections 3.1 and 3.2. The process starts with the double gear which represents the gear, and it is controlled by the shank rotate and hypoid port; these serve as the first system (**Figure 4**). The second system which is the head friction loop is connected by a double-headed screw to the head torque gear rotate and the curve torque that controls the ball drive screw; this serve has the third system with a double-port connection to the rRR1. It initializes the process all over again with a single-port connection to the double-headed gears. This loop is continued till the mated parts are tightly connected. This process is better visualized on the virtual environment created in simulation X.

3. Method

The method applied in this chapter is based on applying a simple algorithm using the system-of-system approach, which helps in determining the effections of tension on the torque, friction on the drive screw, which is described in the proceeding sections. To achieve this, more controls are attached to the threaded screw to understand is mechanics. The torque applied to a fastener is absorbed in different areas, such as the underhead friction, which absorbs almost 50% of the total torque. The thread friction absorbs close to 40% of the applied torque. The remaining 10% of the applied torque represents the clamping force that holds the components together. An increase in friction of say 5% reduces the tension that equals half power. The area under the curve on the torque angle represents the total energy required to tighten a fastener. Some parts of this area represent the clamping force or energy under the bolt and clamped component in the force deformation process. A starting point for all threaded fastener analysis depends on the basic torque-tension equation which is given as

$$T = K \times D \times F \tag{1}$$

This represents the relative magnitudes of torque and clamp force or friction. This also defines a linear correlation between the torque and tension on a bolted screw. The design model is given as

$$T = KdF \tag{2}$$

where T, torque (N m); d, nominal diameter (m); F, force (N); and D, nut factor.

Geometrically, the deformation (δ) of the fastener and angle of turn are related by the following:

$$P_{thread} = \delta \times \frac{360}{\alpha} \tag{3}$$

where δ, deformation (m); α, angle (O^c); and P_{thread} pitch thread (in).

3.1 Tension on fasteners

When the fastener is tightened on a joint infinite stiffness, the relationship correlates directly with the stress induced by the strain in the fastener. The tension produced is totally proportional to the angle of turn from the elastic origin when extensive testing is conducted on the tension product. By projecting the tangent line, the elastic origin can be located near the elastic portion of the torque-angle curve towards the zero-torque zone. The basic correlation of the stress on strain in the elastic region is given as

$$S_{stress} = S_{strain} \times \xi \tag{4}$$

where S_{stress}, stress (in/in); S_{strain}(M/m), strain; and ξ young modulus (N/m^2).

The stretch ∇ or friction of the bolt loaded in tension is computed as

$$\nabla = \delta \times \frac{360}{\alpha} \tag{5}$$

Based on the above given equations, the process for determining the friction on the threaded screw is illustrated in the algorithm (**Figure 4**). This is used to define and describe the mechanics of the given model in **Figure 5**.

The input torque can be thought of as the amount of work (or energy) which is applied to a threaded fastener that causes the bolt to turn in an "α" amount of degrees and in turn stretching the bolt to produce a clamp load. The input torque is added in two ways on the thread and headed torque. The first is the reaction to the input torque and the amount of the torque it takes to keep the nut from turning when an input torque is applied to the bolt head, while the second is the amount of the input torque it carries to overcome friction in the bearing surface of the part of the fastener being turned. To obtain a friction effect M_a, there is a need to combine the force of gravity M_g and Kenal effect M_K. Therefore the input torque is given as

$$M_a = M_g + M_K \tag{6}$$

3.2 The torque effect

The thread torque is a combination of two factors which are thread friction torque and the pitch torque. The frictional portion of the thread torque describing the amount of torque it takes to overcome the friction in the engaged threads and the geometrical faction for the thread torque describing the amount of torque that is used to stretch the bolt.

The graph in **Figure 6** indicates the thread head friction torque which is calculated by subtracting the pitch torque from the thread torque. The graphical representation makes the input torque definition look like an increase in shank transmission on the friction plate. The amount of thread torque and underhead torque presents a shank tightening which can be described using friction coefficients. These are always described in terms of two kinds of material, the shank plating and coatings. It describes the friction between mating threads. The value describing the friction between the bearing materials of the bolt and the first clamping plate and nuts depends on the end of the fasteners which are turned. Also the value describing the overall friction in the joint that combines the aspects of both the bearing surfaces and the threaded head.

Figure 7 indicates the result in friction coefficient based on the plate contact, drive screw, and shank transmission; these are calculated values using the measured input torque, the clamp load, and thread torque alongside some geometric characteristics such as the thread pitch and bearing diameter. A 150 to 100% of the torque input is used to overcome the friction in the threaded and underthreaded regions, with a remainder of the 50 to 0.0% of the input torque left to stretch the bolt producing clamp load. The basic distribution of both the threaded and underthreaded

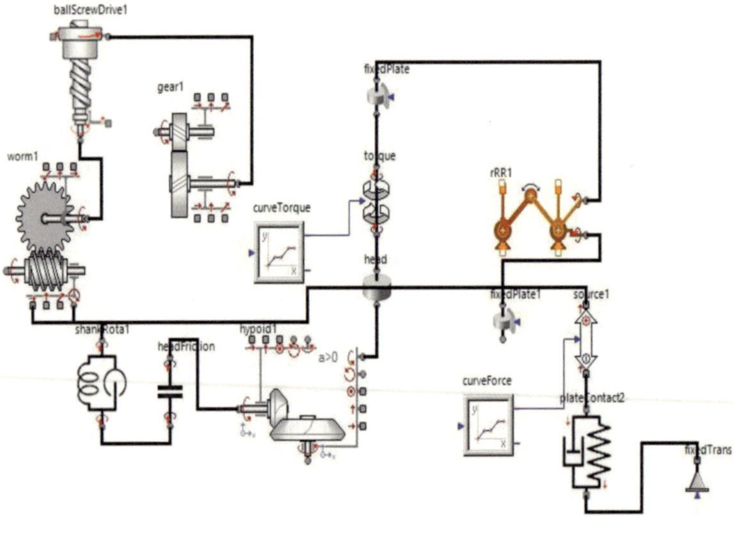

Figure 5.
Determining the friction of the drive screw model with curve torque.

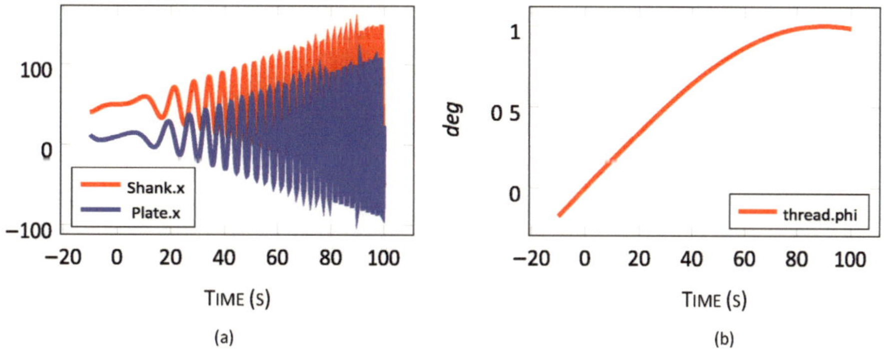

Figure 6.
Rotation angle of the screw and displacement of the screw. (a) Friction behavior on screw drive, (b) Thread displacement on screw.

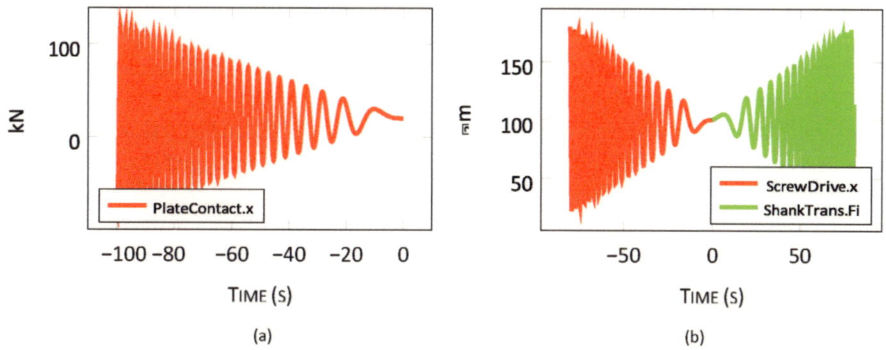

Figure 7.
Forces and deformation of the contacts between screw plate. (a) Force on Plate contact, (b) Deformation on screw drive.

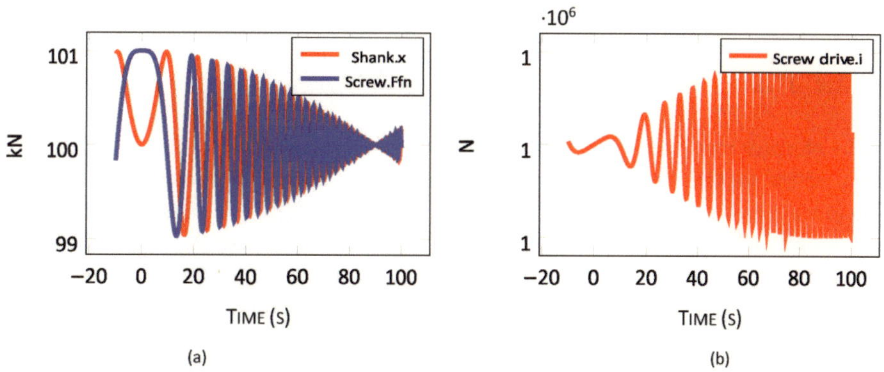

Figure 8.
Friction in the thread and normal force on drive screw. (a) Friction on shank and screw drive, (b) Force on screw drive.

area takes more torque to overcome the friction depending on the bearing surface area along with the friction coefficient between contact surfaces.

Comparing the shank heat transfer and drive screw regarding friction coefficient in **Figure 8**, the wider it takes the friction to get in range, the wider the clamp load range will be at a given torque value for both drive screw and the shank heat. Most tension on drive screw requires tightening a specified bolt/washer

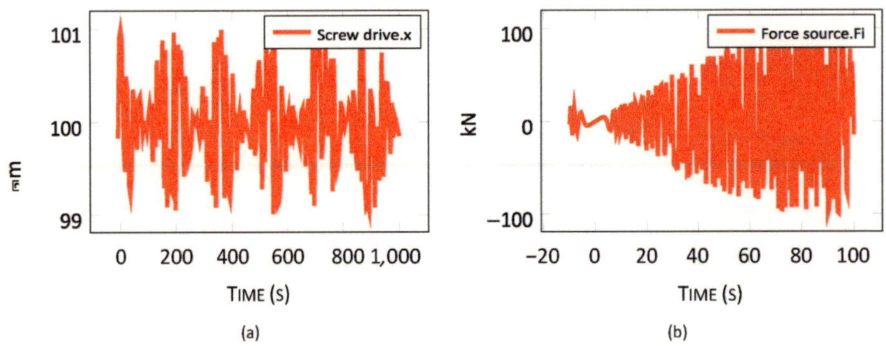

Figure 9.
Motion in the thread and the external force. (a) Force motion on screw drive, (b) Behavior of Force on screw drive

combination to a given clamp load value and returning a torque value. Both the nut and washer have a typical plating and dimensional requirements, and the bolt has requirements for the material and dimensions, which reduces all variances in the joint to the bolt coating for a drive screw.

Figure 9 compares the drive screw and force on bolt; the input values were being tested at a specified tension value. The coefficient of friction values can be calculated using the default values and Eq. 5 described above along with the torque-tension data. The calculated data is plotted and compared with the source of the force on the bolt or drive screw for this case. Friction variances affect the tightening of a bolted joint occurring in the bearing area and the threads. The friction variances are always a function of two materials of the mating parts. These two materials are the hypoid and worm effect on the drive screw, which makes it possible for the torque and tension behavior to be synchronous. The torque is a measure of the models' twisting force which is required to spin the nuts along the thread, while the tension allows the stretch of the bolt that provides the clamping force of the joints. So the plating used have a consistent relationship between the torque and tension.

4. Conclusion

The chapter mostly focused and limited to determining the effect of friction on a drive screw for the optimization on advanced mechanical purposes. The torque value effect are strictly based on the friction obtained, which is present in the bolted joint assembly. The sum of the variances in the friction leads to the variance in the amount of tension attained at the specified input torque value. The friction variation is a function of two materials of the mating parts. Likewise the coefficient of the frictions is obtained for all parameter comparisons, which is not given in terms of the material plating. The simulation effect is used to limit the amount of variance allowed in the friction coefficients against a specified material. It allows for the end user to basically be confident that the plating used has a consistent relationship between both torque and tension.

For future purposes, it is recommended that the torque time print of all the torque and audit simulation be recorded and reviewed before specification of the equipment and settings of numerical limits of the manual simulation. When conducting these basic simulations, it should always include the torque angle and its time print or signature of analysis for the installation steps. This is required prior to any meaningful correction factor that is established in relation to installation and audit recordings.

Acknowledgements

The authors wish to acknowledge and thank both the Nasarawa State University and Ahmadu Bello University Zaria, for providing the modeling software and assistance during the production of this chapter.

Author details

Fatima Isiaka[1,2*†], Awwal M. Adamu[2†], Salihu A. Abdulkarim[3†] and Abdullahi Salihu[1†]

1 Department of Computer Science, Nasarawa State University, Nasarawa State, Nigeria

2 Department of Electrical Electronics, Federal University of Technology, Minna, Nigeria

3 Department of Software Engineering, Federal University Dutse, Dutse, Nigeria

*Address all correspondence to: fatima.isiaka@outlook.com

† These authors are contributed equally to this chapter.

References

[1] Available from: https://en.wikipedia.org/w/index.php?title=Bolt-(fastener)andoldid=920539745

[2] Bickford JH. An Introduction to the Design and Behavior of Bolted Joints. New York: Marcel Dekker, Inc.; 1990

[3] Virgil Moring Faires. Design of Machine Elements. 3rd ed. New York: The Macmillan Company; 1955

[4] Blake A. Design of Mechanical Joints. New York: Marcel Dekker, Inc.; 1985

[5] Blake A. What every Engineer Should Know About Threaded Fasteners Materials and Design. New York: Marcel Dekker, Inc.; 1986

[6] Forrest PG. Fatigue of Metals. New York/Reading, Massachusetts: Pergamon Press/Addison-Wesley Publishing Company, Inc.; 1962

[7] Industrial Fasteners Institute, editor. Fastener Standards. 5th ed. Cleveland, Ohio: Industrial Fasteners Institute; 1970

[8] Drumheller J. Fundamentals of torque-tension and coefficient of friction testing. In: White paper No. 21. MTS Systems Corporation; 2019. pp. 1-14

[9] Industrial Fasteners Institute, editor. Metric Fastener Standards. 2nd ed. Cleveland, Ohio: Industrial Fasteners Institute; 1983

[10] Juvinall RC. Engineering Consideration of Stress, Strain, and Strength. New York: McGraw-Hill Book Company; 1967

[11] Juvinall RC. Fundamentals of Machine Component Design. New York: John Wiley and Sons; 1983

[12] Kulak GL, Fisher JW, Struik JHA. Guide to Design Criteria for Bolted and Riveted Joints. 2nd ed. New York: John Wiley and Sons; 1987

[13] Shigley JE. Mechanical Engineering Design. New York: McGraw-Hill Book Company; 1977

[14] Shoberg RS. In: Director of Technology, editor. Engineering Fundamentals of Threaded Fastener Design and Analysis. PCB Load and Torque, Inc; 2019. pp. 1-39

Chapter 8

A System-of-Systems Perspective on Frequency Estimation: Time-Frequency Distribution of Multiple LFM Signals

Ruolun Liu, Xueqin Zhang and Rui Huang

Abstract

This chapter provides a System-of-Systems (SoS) perspective on a study of frequency estimation of signals with a focus on Linear Frequency Modulation (LFM) signals. This chapter describes an SoS approach for frequency estimation using Chirplet Transform (CT), Hough Transform (HT), and the Short Time Fourier Transform (STFT) with filtering viewpoint. The filtering viewpoint employs the filter impulse response length to obtain the best time-frequency concentration for accurate estimation of a signal frequency. The optimum impulse response length can be found by varying the length of the filter impulse response and observe the changing in the time-frequency distribution (TFD). The chapter shows that when the length of the impulse response becomes longer, the time-frequency concentration in TFD increases first and then decreases.

Keywords: LFM Signal, TFD, STFT, filter bank, SoS

1. Introduction

The study of stationary deterministic signal has been greatly explored and appreciated. On the other hand, most signals encountered in applications are random and nonstationary. Unlike the time-invariant statistical properties of stationary signal, the statistical properties of nonstationary signal are normally time-variant where a time-frequency combined analysis tool, time-frequency distribution (TFD), is required to observe the nonstationary signal in the time domain and the frequency domain at the same time. The early TFD is often given by Short Time Fourier Transform (STFT), Gabor Transform, or Continuous Wavelet Transform (CWT). The classical one is the quadratic Wigner-Ville Distribution (WVD). The latest type would be the parameterized TFD [1–3] developed in recent years. Both STFT and CWT, in the sense of transform, are not able to achieve a fine resolution in both time and frequency domain simultaneously, due to the restriction of the Heisenberg–Gabor inequality. Linear frequency modulated (LFM) is one of the pulse compression techniques in the Radar system to solve the conflict between rang and resolution where the carrier frequency is continuously modulated during the pulse duty time. The term of instantaneous frequency (IF) is used to describe how the carrier start frequency changes linearly all the way up to the end frequency.

In fact, the LFM signal has no fixed period nor frequency within each pulse duty time. The quadratic WVD will achieve the highly accurate frequency component for noise-free LFM signal, where the constant amplitude brings WVD a row of delta functions along the linear IF trajectory [4, 5]. In the case of noisy LMF, the WVD peak position will bias from the true IF, where the bias-to-variance tradeoff is inevitable in the IF estimation. The Chirplet Transform (CT) [1, 2] is a typical parametric TFD, which is particularly designed for the analysis of chirp-like signals with linear IF. In the initialization process of the CT, the parameters estimation is based on the peak of the STFT magnitude, thus the estimation results are greatly affected by the background noise. For the multiple LFM signal, it is difficult to distinguish and track multiple IF lines. If the Hough Transform (HT) is applied to the spectrogram magnitude first, the robust parameter estimation can be obtained for each component, a set of time-frequency images can then be emerged by post processing to finally get a TFD with higher concentration.

Some scholars have analyzed the nonstationary signal with filtering viewpoint. In order to analyze the audio signal, Brown proposed the constant Q transform (CQT) [6], where the central frequency of each band is not uniformly distributed and its frequency resolution is not a fixed value in the frequency domain, that is more suitable for nonstationary audio signal processing. Another adaptive filter bank is proposed in [7] where the frequency resolution is changed by adjusting the window length in each sub-band. A more generalized TFD is proposes based on the traditional CQT [8]. It can be used to define a time-frequency analysis framework with arbitrary central frequency at arbitrary frequency resolution. The parameters in the framework are clearly defined to achieve a good resolution at any given frequency range. Another novel time-frequency analysis is proposed in [9] where the filter bank is a high-resolution Gaussian filter bank. Based on the nonlinear characteristics of the human auditory system, the Gaussian filter bank is designed to adjust the central frequency of each band. At the same time, the multi-resolution characteristic of the filter bank is discussed based on the idea of Wavelet transform.

However, little has been reported about the STFT in the filtering viewpoint. The STFT is always regarded as time shifted Fourier transformations where frame length is fixed for every transformation. If STFT is treated as the outputs of a filter bank, the length of each channel impulse response can be set differently according to the different signal frequency of different channel. This chapter combines our two recent conference papers [10, 11] and provides a unified framework derived from a System-of-Systems perspective for analyzing LFM signal using Filtering Viewpoint approach. Based on a series of experiments, the impact of the filter impulse response length is observed with the variation of TFD. It is also proved that longer filter does not always guarantee a better energy concentration in the TFD. To obtain the best TFD for the LFM signal processing, an optimal impulse response length needs to be determined beforehand.

This chapter presents a SoS approach for frequency estimation with a focus on LFM signal. Using a standard system engineering approach, we can decompose the frequency estimation process into three systems consisting of:

- System 1: responsible for transforming the time-domain signal into (i) frequency-domain signal, and (ii) TF transform using CWT. For LFM signal type, CT is selected for CWT. Section 2 provides detailed description of STFT and CT transforms.

- System 2: responsible for detecting instantaneous signal frequency (IF) in the presence of noise. Section 3 describes a proposed technique using Hough transform for detecting IF straight lines.

- System 3: responsible for (i) assessing of filter impulse response length on TFD (see Section 4), and (ii) analyzing the time-frequency behavior and applying Hough transform for frequency estimation (see Section 5).

Section 4 presents a series of experiments showing the influence of the impulse response length to the time-frequency concentration in TFD and provides the steps of finding the optimal impulse response length. Section 5 provides an example of time-frequency analysis and proves the feasibilities of the proposed approach before the conclusion.

2. Frequency transforms: a review

2.1 STFT using filter bank viewpoint

The STFT is normally regarded as the Fourier transform of the framed signals with an observation window of fixed length. Whereas being looked from the filtering viewpoint, the filter bank has some advantages that the traditional transform does not have. For example, if the frequency range of the signal is known, the corresponding bands can be selected in advance and only the selected bands need to be calculated, which will greatly reduce the computation cost. From the viewpoint of filtering, STFT can be actually regarded as the outputs of a filter bank, in which each band is called "analytical filter" and has its own impulse response function, $w[n]$. The STFT at the band centered at ω_0 is the output of the analytical filter driven by the signal demodulated with the complex exponential carrier at the central frequency of that band.

$$X(n, \omega_0) = \sum_{m=-\infty}^{\infty} \left(x[m]e^{-j\omega_0 m}\right)w[n-m] \tag{1}$$

Rewrite the demodulated signal into $x_w[n]$, the STFT at ω_0 has no difference to the discrete convolution between the driven signal $x_w[n]$ and impulse response $w[n]$.

$$X(n, \omega_0) = x_w[n] * w[n] \tag{2}$$

This can also be expressed as:

$$X(n, \omega_0) = e^{-j\omega_0 n}\left(x[n] * w[n]e^{j\omega_0 n}\right) \tag{3}$$

where the sequence $x[n]$ passes a filter with the modulated impulse response first, before getting demodulated with the same carrier.

Following this idea, the entire STFT covering N discrete DFT frequencies can be obtained as the output of the filter bank shown in **Figure 1**, where $K = N\text{-}1$, and N is the frame length of STFT. In fact, when this system deployed directly into the hardware, the carrier frequency of each band-pass filter can be arbitrarily selected, which could have nothing to do with N. Another benefit of the direct deployment is the time resolution of this filter bank output, which could be as high as the time resolution of the input signal $x[n]$. The challenge is the selection of response time of each selected band may be different from band to band because of the different filter lengths are required to provide high time-frequency concentration in the filter bank output $x[n, k]$.

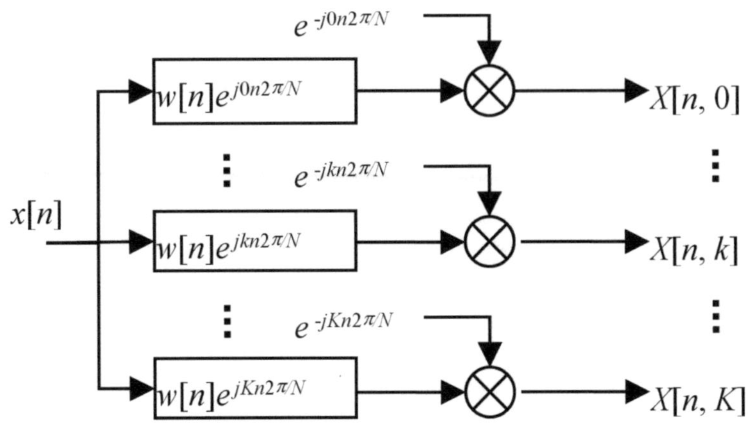

Figure 1.
STFT at N DFT frequencies using filtering viewpoint.

2.2 Chirplet transform (CT)

The CT of a signal $s(t)$ is defined as

$$CT_s(t_0, \omega, \alpha; \sigma) = \int_{-\infty}^{\infty} z(t)\psi^*_{t_0,\alpha,\sigma}(t) \exp(-j\omega t)dt \tag{4}$$

where $z(t)$ is the analytical signal of $s(t)$ generated by the Hilbert transform, and ψ is a complex window given by

$$\psi_{t_0,\alpha,\sigma}(t) = w_\sigma(t - t_0) \exp\left[-j\frac{\alpha}{2}(t - t_0)^2\right] \tag{5}$$

where t_0 is the time shift and α the chirp rate. The window function w is usually taken as Gaussian function expressed as

$$w_\sigma(t) = \frac{1}{\sigma\sqrt{2\pi}} \exp\left[-\frac{1}{2}\left(\frac{t}{\sigma}\right)^2\right] \tag{6}$$

From this definition, it can be seen that the CT can be decomposed into a series of operations: 1) rotating the signal under consideration by an angle in the time-frequency plane; 2) shifting the signal by a frequency increment; and 3) applying STFT with the Gaussian window.

This process can be depicted in the **Figure 2**. The solid line is the IF line of the target LFM signal that has the IF function $\omega(t) = \omega_0 + \lambda_0 t$. The dot-dashed line is the IF line after the rotation, and the dashed line represents the IF line after translation.

As mentioned earlier, given a set of properly determined kernel characteristic parameters, the CT could produce a high-quality TFD for a considered signal. The result can have an excellent T-F concentration, which measure the IF trajectory width over the TFD surface, so the IF trajectory can be easily identified. Therefore, the determination of proper parameters is critical for the application of the CT method. Briefly speaking, the basic idea of the CT based T-F analysis uses the kernel characteristic parameter ($\alpha = 0$) to form the TFD, and then finds the maximum value along time axis in the time-frequency plane. The resulting maximum line approximation is considered to be an IF trajectory. The chirp parameter obtained by the line fitting will be reapplied to the CT transformation. The procedure can be

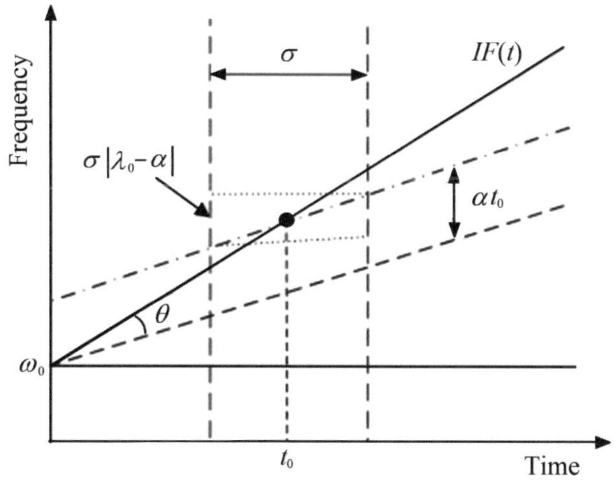

Figure 2.
The Chirplet transform.

repeated until no evident improvement is observed in the T-F concentration of the TFD.

To measure the T-F concentration of the TFD, the Rényi entropy can be used with the definition:

$$E(s) = -\iint \log |CT_s(t,\omega)|^3 dt d\omega \tag{7}$$

And the termination condition can be set as

$$\xi_s = |E_{i+1}(s) - E_i(s)| < \delta \tag{8}$$

or

$$\xi_s = \frac{|E_{i+1}(s) - E_i(s)|}{|E_{i+1}(s)|} < \delta \tag{9}$$

where δ is a predetermined threshold. If the parameters of the initialization are not accurate, it will lead to be a lower overall T-F concentration. Because STFT is susceptible to noise, the robustness of the whole algorithm is not strong. Fortunately, the fractional Fourier transform can compensate for this shortcoming.

3. Detect IF straight lines by Hough transform

Through the description of the previous section, we know that as long as we can get the precise FM parameters, we can get the TFD with high T-F concentration. However, for multicomponent signal with low SNR, the difficulty will be significantly huge. Inspired by the reference [12], we apply the robust HT to detect the IF lines in the CT-TFD to depress the noise during the process of line fitting. The reason why we adopt this image processing technique is that the HT can detect multiple lines accurately, even in the low SNR situations.

For an IF straight line in the Cartesian coordinate plane, there are two common representations: point-slope form and two points form. In the HT, however, another representation is considered: coordinate (r, θ) is used to represent a straight

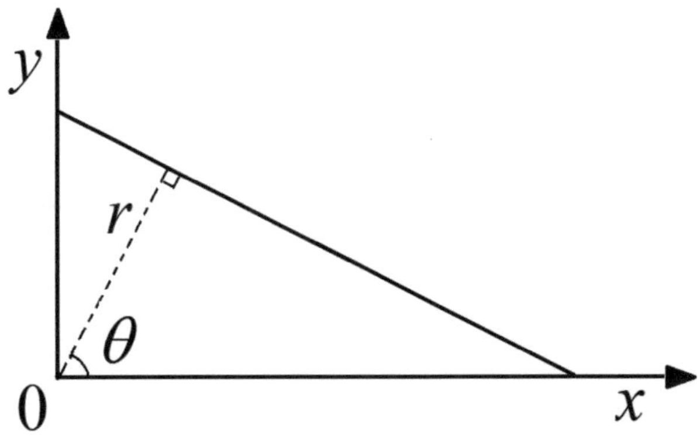

Figure 3.
The line representation in Hough transform.

line where r is the distance from the straight line to the origin, θ is the angle between the x-axis and the perpendicular line passing through the origin as shown in **Figure 3**.

The idea of using HT to detect straight lines needs to assume M straight lines for each point, usually $M = 180$. In this chapter, the angle resolution of the detected line is 1 degree. The coordinates (r, θ) of the M lines are calculated respectively. If there are a total of L points to be checked, the total numbers of coordinates are $M \times L$. If a number of points are on the certain same line, then there must be a same number of points with the same the same coordinate (r_c, θ_c). The following example shows that if there are three points in the plane, we can determine whether or not the three points are on the same straight line by their HT coordinates.

4. Impact of filter impulse response length on TFD: simulation results

The below simulations are all conducted on the Matlab 2015b installed in Windows10 system on the Dell T7910 workstation with 2 Intel Xeon E5-2630v3 CPUs and 256G LRDIMM memories.

To see the influence of the filter impulse response length on the concentration of the TFD, the LFM signals of different FM parameters start with the simulated signal:

$$z_1(t) = \exp\left[j(12\pi t + 10\pi)t\right] \tag{10}$$

where its IF line function is $f_1(t) = 6t + 5$. The signal lasts for 5 seconds, and 1000 channels are selected uniformly from $0 \sim 50$ Hz. The sampling frequency of $z_1(t)$ is fixed at 100 Hz, but the impulse response length of all the bands is tested on 40, 80, 100 and 150 samples respectively. The corresponding TFDs are shown by **Figure 4**.

It can be seen clearly that as the impulse response length grows, the T-F concentration of the TFD increases first and then decreases. In order to check whether this is a universal phenomenon, the following multiple LFM signal is built.

$$z_2(t) = \sum_{k=0}^{7} \exp\left[j(2k+1)5\pi t^2 + 10\pi t\right] \tag{11}$$

The IF slopes of all the components of $z_2(t)$ form a sequence of common difference of 5 Hz, starts from 2.5 Hz to the highest 37.5 Hz. Thus, the IF of $z_2(t)$ is in the range of $5 \sim 192.5$ Hz within the 5 s duration. In the **Figure 5**, the TFDs are given for the impulse response length of 50, 100, 150, 200, 250, 300, 350, 400, 450, 500, 550

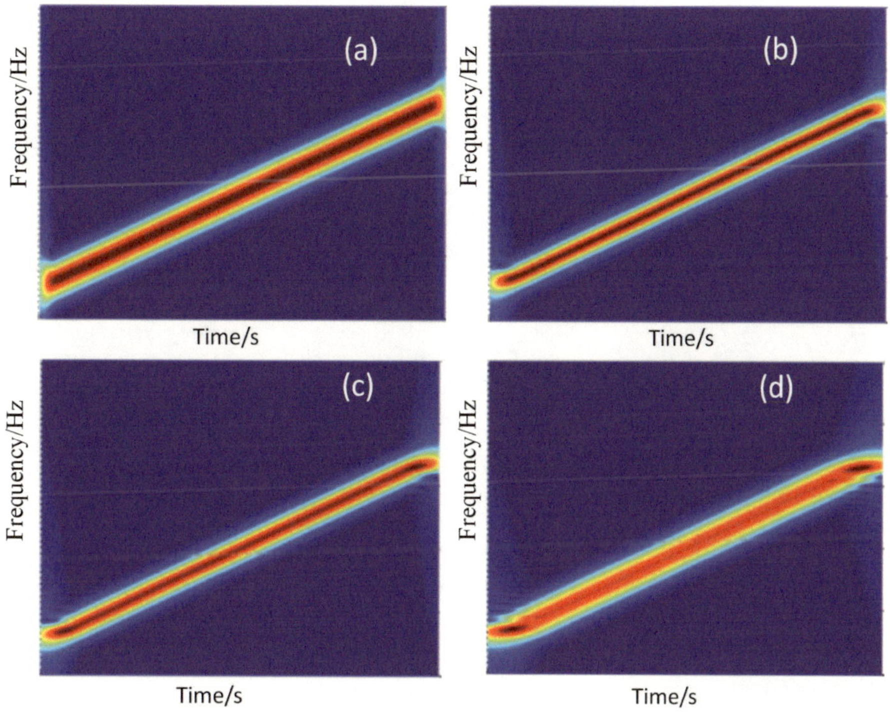

Figure 4.
TFDs of $z_1(t)$ using filter bank at 4 different filter lengths of 40 (a), 80 (b), 100(c), and 150 (d) samples.

and 600 samples respectively at the sampling frequency of 1000 Hz. 300 bands are uniformly selected from $0 \sim 300$ Hz in the calculation of each TFD. It is shown clearly again by the **Figure 5**, as the impulse response length increases, the T-F concentration of each component in the signal $z_2(t)$ increases first and then decreases. The larger the frequency modulation coefficient (or IF slope) is, the faster its TFD reaches its best time frequency concentration. For the component with the lowest IF slope of 2.5 Hz/s, the concentration change is not obvious since it is close to a stationary signal.

In order to verify the above conclusion, a complex stationary signal below of 5 Hz is also tested.

$$z_3(t) = \exp(j10\pi t) \tag{12}$$

The testing range of the impulse response length, the selected channels, the signal duration, and the sampling frequency are taken in the same way as that of $z_1(t)$. The resulted TFDs are shown in the **Figure 6**. It can be seen that for this stationary signal $z_3(t)$, as the impulse response length increases, the T-F concentration of the TFD keeps increasing because longer filter collects nothing but more energy of the stable signal, which is different from the case of LFM signal where more interferences will be observed by the longer filter. This is to say that the length of the impulse response should match the changing speed of the signal IF, so that the TFD could accurately reflect the T-F trajectories of the LFM signals.

The next simulation is about the harmonically related multiple LFM signals as given below.

$$z_4(t) = \sum_{k=0}^{5} \exp\left[j6\pi t^2 + 20(2k+1)\pi t\right] \tag{13}$$

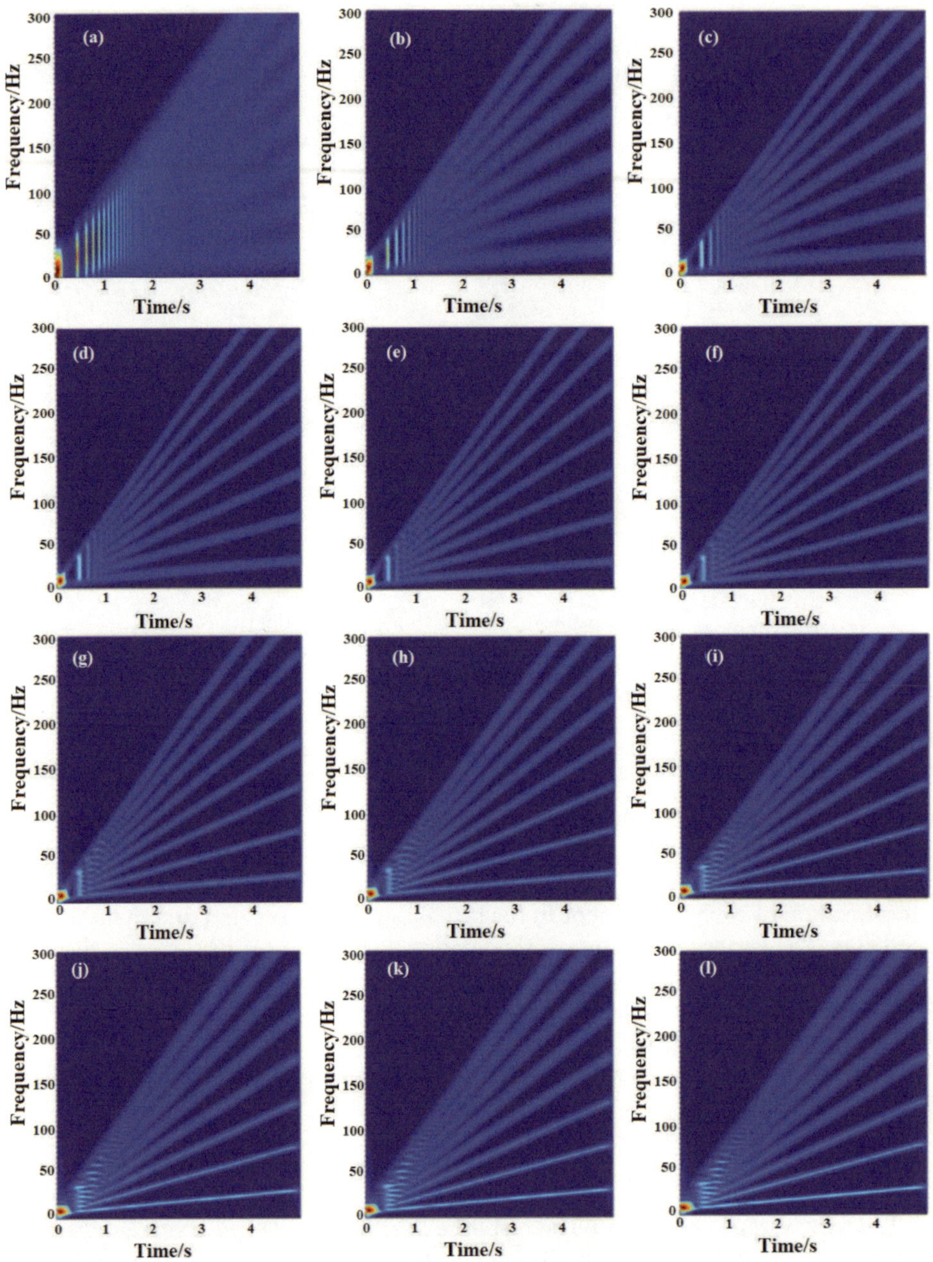

Figure 5.
TFDs of $z_2(t)$ using filter bank with 12 different filter lengths running from 50 (a) to 600 (l) samples at the step of 50 samples.

The length of the impulse response runs within {150, 250, 350, 450, 550, 650}. The selected channels, the signal duration, and the sampling frequency are taken in the same way as what has been done on $z_2(t)$. The resulted TFDs are shown in the **Figure 7**.

It can be seen from **Figure 7** that with the increasing of impulse response length, the TFD T-F concentration of each component of the signal $z_4(t)$ gets better and better, while the T-F concentrations of different components with the same filter length have no difference from each other. In other words, the same IF slope results the same TFD T-F concentration given the same filter length.

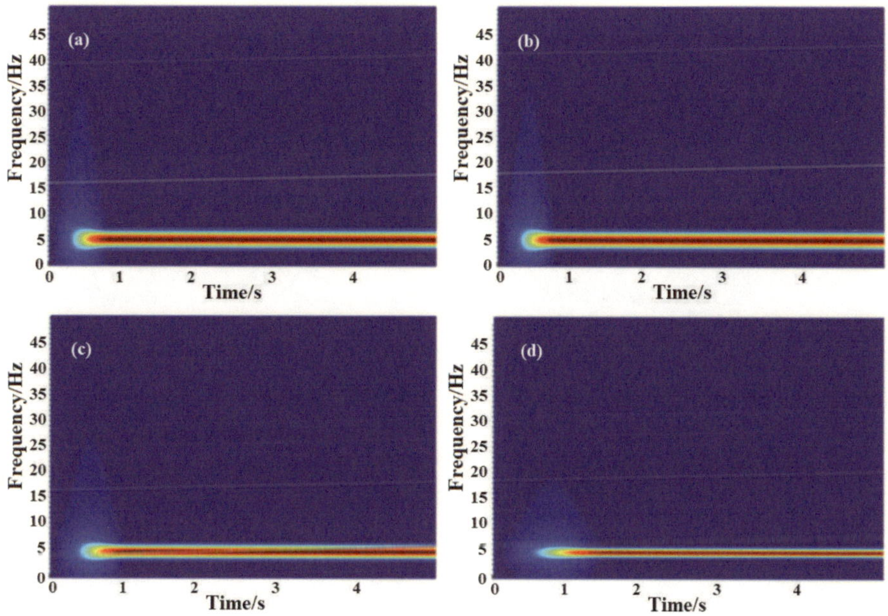

Figure 6.
TFDs of $z_3(t)$ using filter bank at 4 different filter lengths of 40 (a), 80 (b), 100(c), and 150 (d) samples.

Through the description in the previous section, the high T-F concentration over the TFD surface can be obtained if the impulse response length matches to the IF slope of FLM signal. In order to obtain this optimal filter length, the T-F concentration $M(l)$ is measured by the averaged bandwidth in the TFD as given below, where $A_l(t,\omega)$ is the TFD magnitude calculated with the filter length of l at the (t,ω) point, T is the time limit of TFD. Based on this TFD concentration measurement, the optimal filter length can be determined by Eq. (15). The IF of the signal of interests can then be accurately estimated from the resulted $A_{L\text{opt}}(t,\omega)$ using any modest ridge detection method.

$$
\begin{aligned}
M(l) &= \frac{1}{T}\sum_{t=1}^{T}\sum_{\omega} B_l(t,\omega) \\
B_l(t,\omega) &= \begin{cases} 1, & A_l(t,\omega) > 0.9 \max\limits_{\arg\omega}[A_l(t,\omega)] \\ 0, & otherwise \end{cases}
\end{aligned}
\tag{14}
$$

$$
L_{opt} = \arg\max_{l}[M(l)] \tag{15}
$$

When the LFM slope coefficient is given, the channels of the filter bank can be precisely selected to cover that IF range. Even for the signal with unknown LFM slope coefficient, one can always observe the rough IF range using traditional STFT, so no difficulty will be met in the filter bank channel selection. **Figure 8** compares the proposed TFD of filtering viewpoint with the traditional STFT of transform viewpoint. The LFM signal is $z_1(t)$ sampled at the frequency of 100 Hz, the impulse response length is 80 samples for all the channels, the selected 1000 channels uniformly divide the range of $0 \sim 50$ Hz, which calls for the frame length of 2000 samples in the traditional STFT to reach the same frequency resolution. It can be seen that the concentration of the proposed TFD is significantly better than that of the traditional STFT.

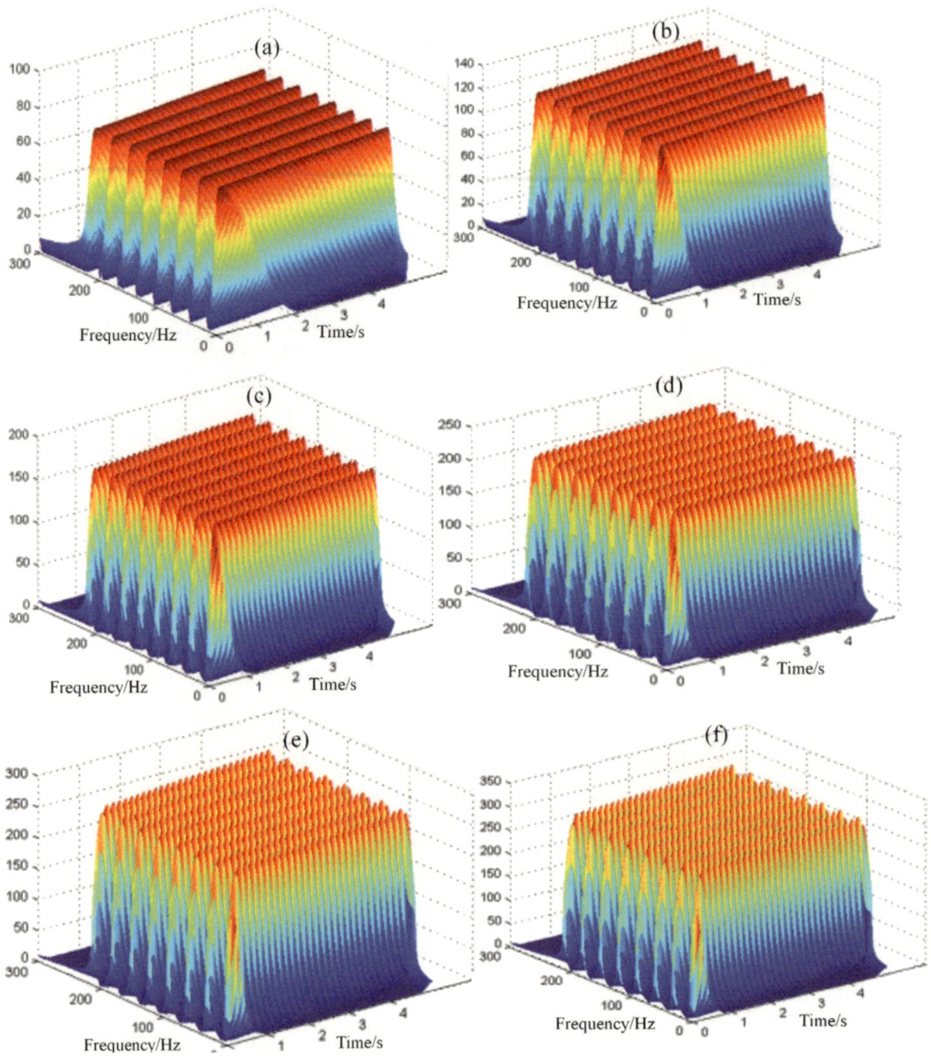

Figure 7.
TFD magnitudes of $z_4(t)$ using filter bank at the filter length of 150 (a), 250 (b), 350 (c), 450 (d), 550 (e), and 650 (f) samples.

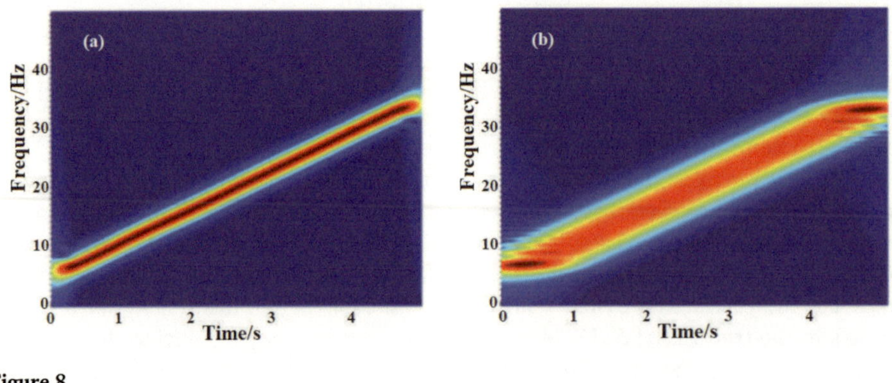

Figure 8.
Proposed TFD (a) vs. traditional STFT (b) of $z_1(t)$ at the same frequency resolution.

5. Frequency estimation: time-frequency analysis for multiple LFM signal based on the CT and HT

Taking the two-component signal as an example, it will give more accurate parameters through line fitting before applying them to CT. In order to see the TFD difference between parameter matched and nonmatched CTs, a simulated multiple LFM signal is considered as:

$$s(t) = \sin\left[2\pi\left(40t - t^2\right)\right] + \sin\left[2\pi\left(10t + \frac{5}{4}t^2\right)\right] \qquad (0 \le t \le 15s) \qquad (16)$$

where IF lines are $f_1(t) = 40\text{-}2\,t$ and $f_2(t) = 10 + 2.5\,t$. So, the actual parameters should be $\alpha_1 = -4\pi$ and $\alpha_2 = 5\pi$. Suppose we have these two exact parameters, and then apply them into the CT transformation and observe the properties of the TFD graphs.

(a) $\alpha_1 = -4\pi$, (b) $\alpha_2 = 5\pi$, (c) the superposition of (a) and (b).

From **Figure 9**, one can see that when the parameter $\alpha = \alpha_1$, the component with IF of $f_1(t)$ has better T-F concentration in the CT-TFD while the component with IF of $f_2(t)$ has very low T-F concentration and also very low magnitude. It shows the opposite situation when $\alpha = \alpha_2$. Adding the two spectra together, a good TFD can be obtained as shown above. If we cut off the very low magnitude with a threshold before adding up the two spectra, a sharper TFD with higher T-F concentration can be obtained as shown by **Figure 10**.

In order to highlight the advantages of HT, the analog signal of Eq. (16) is generated at the SNR of 2 dB, 0 dB, and -2 dB. Then the STFT and the

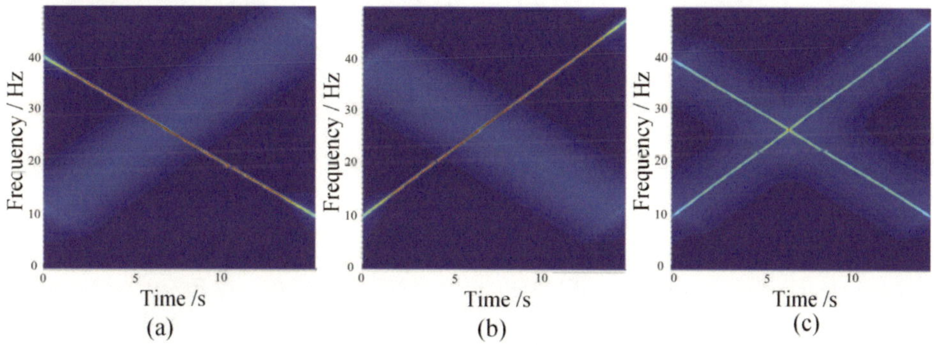

Figure 9.
The TFD results of s(t) using CT with parameter (a) $\alpha_1 = -4\pi$, (b) $\alpha_2 = 5\pi$, (c) the superposition of (a) and (b).

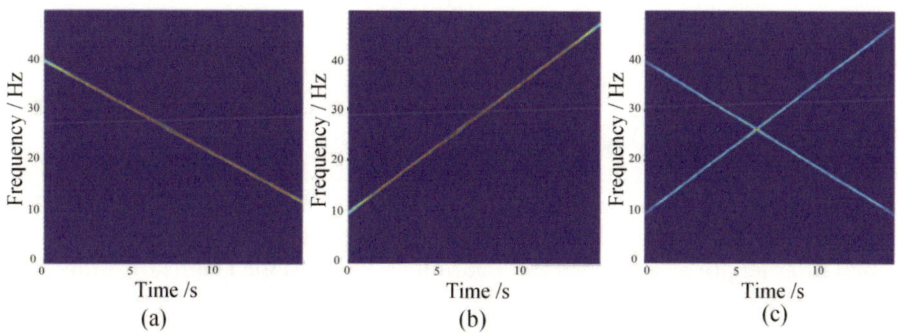

Figure 10.
The results of threshold filtering: (a) $\alpha_1 = -4\pi$, (b) $\alpha_2 = 5\pi$, (c) the superposition of (a) and (b).

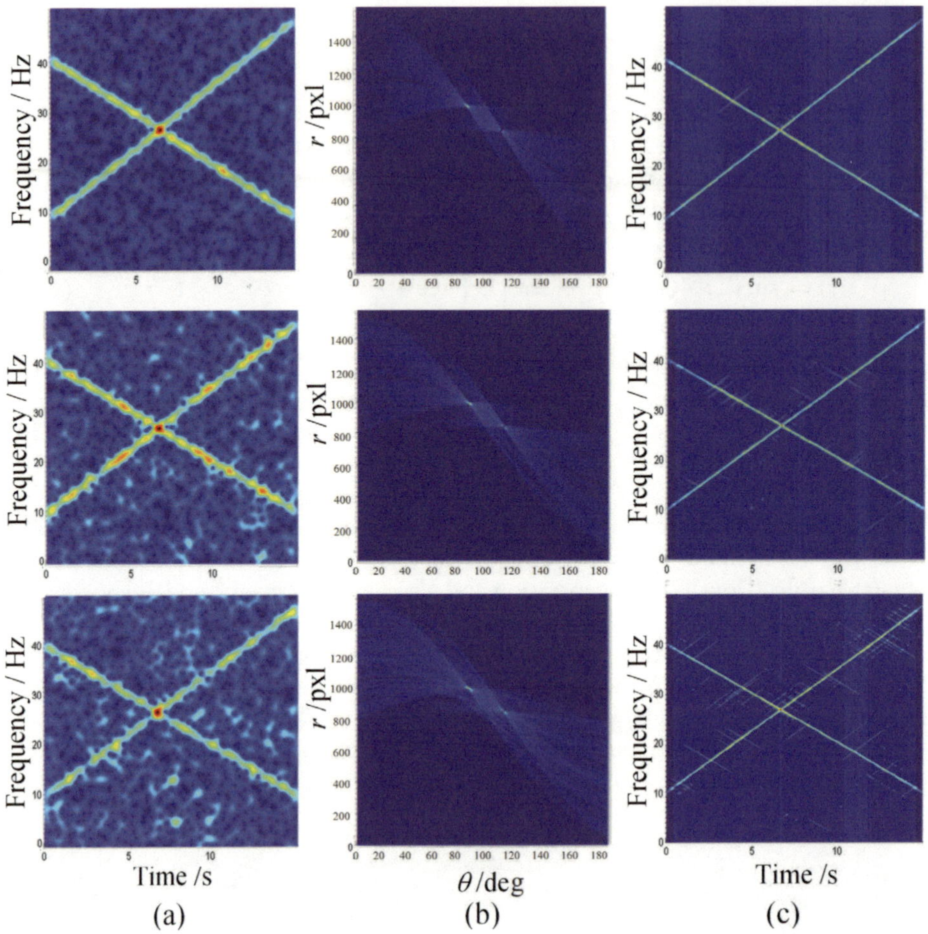

Figure 11.
The STFT (a), the corresponding HT (b), and the TFD after CT & HT (c) of s(t) at the SNR of 2 dB(top), 0 dB(middle), and -2 dB(bottom).

corresponding HT are calculated as shown in **Figure 11**. From the below results, one can see clearly that in the case of low SNR, the STFT is relatively fuzzy, and the ridge edge extraction alone will not give the good results. However, the accuracy of HT is relatively high at all the 3 SNR levels. That is the reason why the HT is adopted in the time-frequency analysis of multiple LFM signal.

6. Conclusions

This chapter investigates an SoS approach for frequency estimation using TFD calculation techniques of the LFM signal through a linear filtering viewpoint. The influence of filter length on the TFD concentration is closely observed through a series of simulations. The simulation results show that the IF slope of the LFM signal is related with the optimal filter length, the higher the slope is, the shorter the optimal filter length is. On the other hand, the same IF slope results the same TFD concentration at the same filter length, no matter how high the IF is. Under the same time-frequency resolution, the traditional STFT shows significantly lower time-frequency concentration than that of the TFD obtained by the proposed filter bank based on the filtering viewpoint. Thought the channel frequency can be freely selected as needed, which saves the computation in the irrelevant frequency bands, many nonstationary signals are not always linearly modulated. For the signals with the nonlinear FM coefficient, the advantage of the filter bank TFD is no longer obvious. This is also a question that needs to be studied further.

The CT has the advantage of high T-F centralization but is easily affected by the noise. In addition, many nonstationary signals are of multicomponent. The decomposition of multicomponent signal [12, 13] into single component signals under noise conditions is a difficult problem. For the multicomponent LFM signal, the Hough transform is adopted to the parametric T-F analysis to obtain the result with good concentration. However, for the multicomponent nonlinear FM signal, there is no effective trajectory detection method, so it will be more difficult to decompose each nonlinear FM component. That will be the focus of our future work.

Acknowledgements

The authors would like to express great thanks to the Shanghai Key Lab in Information Security Management (AGK201709), and Shandong Nature Science Fund Committee (ZR2016FM44) for their financial supports.

Author details

Ruolun Liu*, Xueqin Zhang and Rui Huang
Digital Audio Laboratory, Shandong University, Weihai, China

*Address all correspondence to: ruolun.liu@sdu.edu.cn

References

[1] S.Mann and S.Haykin, "The Chirplet transform: Physical considerations," IEEE Trans. Signal Process., 1995; 43 (11): 2745–2761

[2] Z. K. Peng, G. Meng, F. L. Chu, Z. Q. Lang, W. M. Zhang, and Y. Yang, "Polynomial chirplet transform with application to instantaneous frequency estimation," IEEE Trans. Instrum. Meas., 2100; 60(9): 3222–3229

[3] Y. Yang, Z. K. Peng, G. Meng, and W. M. Zhang, "Spline-kernelled Chirplet transform for the analysis of signals with time-varying frequency and its application," IEEE Trans. Ind. Electron., 2012; 59(3): 1612–1621

[4] J. A. Rosero, L. Romeral, J. A. Ortega, and E. Rosero, "Short-circuit detection by means of empirical mode decomposition and Wigner–Ville distribution for PMSM running under dynamic condition," IEEE Trans. Ind. Electron., 2009; 56(11): 4534–4547

[5] V. Climente-Alarcon, J. A. Antonino-Daviu, M. Riera-Guasp, and M. Vlcek, "Induction motor diagnosis by advanced notch FIR filters and The Wigner–Ville distribution," IEEE Trans. Ind. Electron., 2014; 61(8): 4217–4227

[6] Judith C Brown, "Calculation of a constant Q spectral transform," Journal of the Acoustical Society of America, 1991; 89(1): 425–434

[7] Karin Dressler, "Sinusoidal extraction using an efficient implementation of a multi-resolution FFT," In: Proceedings of the 9th Int. Conference on Digital Audio Effects (DAFx-06), 18-20 September 2006; Montreal, Canada, John Wiley & Sons, 2011. p. 247–252

[8] Thomas Fillon, Jacques Prado, "A flexible multi-resolution time-frequency analysis framework for audio signals," In: Proceedings of the 11th International Conference on Information Science on Signal Processing and their Application, Paris, France, 2012; 3: 1125-1129

[9] Haifeng Zhan, Hongxin Tian, Bo Niu, Conglin Li, "Time-Frequency Analysis Method based on Multi-Resolution Gaussian Filter Bank," J. of China Electronic Science Research Institute, 2017; 12(6): 655-661.

[10] Xueqin Zhang, Ruolun Liu, "Analysis of Linear FM Signal Based on the STFT in the Filtering Viewpoint," 2018 IEEE 3rd International Conference on Signal and Image Processing (ICSIP2018); Shenzhen, China, 23-25 July 2018; p. 389-392

[11] Xueqin Zhang, Ruolun Liu, "Time-frequency Analysis of Multicomponent LFM signal based on Hough and Chirplet Transform," MATEC Web of Conferences 2018;173. DOI: https://doi.org/10.1051/matecconf/201817303054

[12] LJ. Stankovi, M. Dakovi, T. Thayaparan, V. Popovi-Bugarin, "Inverse Radon Transform Based Micro-Doppler Analysis from a Reduced Set of Observations," IEEE Trans. on Aerospace and Electronic Systems, 2015; 51(2): 1155–1169

[13] M. Dakovi, LJ. Stankovi, "Estimation of sinusoidally modulated signal parameters based on the inverse Radon transform," ISPA 2013; Trieste, Italy, 4-6 September 2013; p. 302-307

Section 5

Decision Support Systems (DSS) Applications

Chapter 9

Using Multi-Criteria Optimization in Decision Support under Risk

Andrzej Łodziński

Abstract

The chapter presents an extension of a previous method for decision support under risk. The decision-making process is modeled by a multi-criteria optimization problem, in which the individual evaluation functions represent the results of decisions in several possible scenarios with associated risks. The decision support method is an interactive decision-making process. The choice is made by solving the problem depending on the control parameters that define the aspirations of the decision maker as well as on an evaluation of the obtained solutions. The decision maker selects a set of parameters representing various risks' impacts that influences a solution, and then he/she evaluates the obtained solution by accepting or rejecting it. In another case, the decision maker selects a new value and the problem is solved again for the new parameter. In this chapter, an example of supporting decision-making under risk is presented.

Keywords: decision under risk, multi-criteria optimization, symmetrically efficient decision, scalarizing function, method of decision selection

1. Introduction

In real decision-making problems, the evaluation of a decision is usually nondeterministic, because each problem concerns future activities and is evaluated in terms of future results. A significant portion of the parameters determining the decision conditions and assessment of the results may change, for example, raw material prices, product prices, currency exchange rates, and the sales potential of a given product.

The paper [1] presents a method of modeling decisions under risk in the form of a multi-criteria optimization problem. In this chapter, this approach is developed to apply multi-criteria optimization to supporting decision-making under risk.

As pointed out in [1], when making a decision, the decision maker must take into account both the choice of decisions and the risk's conditions that may occur in his/her environment. Depending on the degree of knowledge of the decision-making situation (features of the problems being solved and the nature of the environment), decisions can be made in a situation of certainty, uncertainty, or risk. This chapter extends the previous work to represent three types of decision-making under risk. The first type of decision-making: decisions are made under conditions of certainty when the decision maker has accurate and reliable information on which to base his/her actions. The effects of the actions can be predicted with high accuracy. The second type of decision-making: a decision maker has a situation of uncertainty when he can determine what factors will affect the decision-making situation, but he cannot determine the probabilities of their occurrence and therefore also the risks' impacts of the decisions taken. The other type of decision-making is a decision in risk conditions that applies to situations where the decision maker can determine what factors will affect the decision situation and determine the probabilities of their occurrence. The decision maker, using his experience and information from the environment, can determine with known or estimated probability the effects of decisions as well as the circumstances surrounding them.

Decision-making under risk is a process in which the results of actions taken by the decision maker are uncertain due to the potential of unforeseen circumstances, factors interfering with these circumstances, or disruptive factors, for example, ambient conditions, called scenarios. These, in turn, are caused by factors independent of the decision maker and have a significant impact on the results of the decision. Examples of scenarios can be: good or bad weather in the future; decline, stabilization, or rising stock values on the stock exchange in the future; and different price values and order volumes for a company operating in the future. Each such variant is a scenario. At the same time, each scenario clearly defines the implementation of results for individual decisions. Only the past is known from experience; we observe the present and try to predict the future. Such predictions are related to the construction of probable scenarios based on statistical analysis of the past data in order to find indications about the future and to anticipate it as accurately as possible. The decision maker is not able to determine with certainty which actions will lead to a result, but he can calculate the probability that a given result will occur. Specific scenarios correspond to the appropriate implementation of the assessment function. For each scenario, we are interested in the best evaluation value [1, 2].

As pointed out in [1], the theory of decision-making under risk refers to utility function and two-criterion techniques (Markowitz-type models). The utility function of the decision maker ensures complete order. If it is known, then the optimal decision is one that maximizes the expected utility [3–5].

This chapter shows an extension of [1] on how the decision problem under risk can be modeled with the multi-criteria optimization, that is, simultaneous minimization of a vector evaluation function whose particular coordinates represent the result of the decision when the given scenario is under risk occurred. The traditional approach [1] to solving a multi-criteria optimization problem requires the introduction of a single scalar objective function valuating individual y vectors and hence the decision vectors x. The solution of the decision problem is then reduced to determining the solution of the optimal single-criteria optimization problem. This approach implies the assumption that the preference relationship can be described using the utility function, u. The major difficulty in solving multi-criteria decision problems is due to the inability to determine a single aggregate quality indicator a priori, while the utility function is just such an indicator. Multi-criteria optimization techniques allow you to solve such a problem without using utility function models. This provides to the interactive multi-criteria techniques for decision support under risk. There are tools of the interactive analysis to define decision support process. They depend on additional preference information gained interactively from the decision maker, allowing simultaneously the decision maker to learn the problem during the process with possible evolvement of the preferences. The effective decision support is using the reference point method. Using the multi-criteria optimization approach, there is no need to identify the utility function of the decision maker. This approach is good for any decision maker who makes decisions under risk conditions (in a cost problem where less is better) for which less is better. This is consistent with first-order stochastic dominance.

The chapter is organized as follows:

- Section 2 presents a modeling approach of decision under risk.
- Section 3 defines a symmetrically effective decision that resolves the decision problem under risk conditions.
- Section 4 discusses the technique of generating symmetrically effective decisions and the method of supporting the decision maker.
- Section 5 gives an example of the application of the proposed decision support approach to a discrete problem.
- Section 6 provides a conclusion of this chapter.

2. Modeling of decision under risk

This section discusses how multi-criteria optimization methods can be used to model decisions under risk. The problem of multi-criteria optimization in the decision space and in the assessment of decision space is formulated.

Decision-making under risk is modeled by introducing scenarios, which represent possible states of the environment. Scenarios are factors that influence the outcome of a decision but are beyond the influence of the decision maker. For example, the risk factors can be raw material prices, product prices, currency exchange rates, deposit rates, and demand (e.g., sales opportunities for a given product) which may change. There may also be a catastrophic event changing the situation, for example, closing of the sales or supply market (e.g., due to embargo), customer insolvency, loss of license, etc.

The scenarios representing the risk factors are presented according to their probability distribution. If we assume that the probability of each scenario is a rational number, then by repeating relevant scenarios, it is possible to approach a situation where the probability of each scenario is the same, for example, selection between random variables Y' and Y'':

$$P(Y' = x) = \begin{cases} 1/2 & x = 2 \\ 1/2 & x = 4 \end{cases} P(Y'' = x) = \begin{cases} 1/4 & x = 1 \\ 3/4 & x = 5 \end{cases}$$

is equivalent to the problem of choosing between two lotteries $y' = (2,\ 2,\ 4,\ 4)$ and $y'' = (1,\ 1(?),\ 5,\ 5)$ with equally probable outcomes, where the order of outcomes is not important.

The number of occurrences of a specific scenario corresponds to the probability assigned to it. The specific set of scenarios $S_i, i = 1, \dots, m$ corresponds to the appropriate realization environment conditions associated with the evaluation function $f_i(x), i = 1, \dots, m$, where $x \in X_0$, a decision set that belongs to the set of admissible decision. There is an assessment function associated with each scenario. At the same time, each scenario clearly defines the implementation of results for the individual evaluation function. For each scenario, a lesser value of the evaluation function is preferred [4, 6, 7].

They are given as:

- the feasible decision set $X_0 \subset R^n$;
- the set of scenarios $S_1, S_2, \dots, S_m$ and the set of probabilities $p_1, p_2, \dots, p_m$ of occurrence of each scenario, and these probabilities are assumed to be known to the decision maker; and

- the decision assessment function, x, at the scenario $f_i(x), i = 1, ..., m$, where there is one evaluation function, $f_i(x)$, associated with each scenario.

The problem of decision under risk is modeled in the form of some kind of multi-criteria optimization problem:

$$\min_x \{ (f_1(x), ..., f_m(x)) : \quad x \in X_0\} \tag{1}$$

This is a special problem of multi-criteria optimization in the sense that all assessment functions are expressed in the same units. This differs from the standard multi-criteria optimization problem, where evaluation functions can be expressed in different units. In the case of modeling decisions in risk conditions, individual assessments, although generated by different functions, are all expressed on the same scale, which allows comparison of their values.

There are as many assessment functions in a multi-criteria problem as there are scenarios. Each scenario has a different assessment function. You want to have the best score for all scenarios.

In the problem of multi-criteria optimization, all values for all scenarios are taken into account and by not looking at the values in each scenario (and not looking at individual coordinates). The result of the decision is the grade vector. You want to have the best score for all scenarios. One grading vector that gives the best score for all scenarios is sought.

The function, f, assigns to each decision variable vector, $x \in X_0$, an evaluation vector, $\mathrm{y} = \mathrm{f}(x)$, which measures the quality of decisions, x, from the point of view of the determined system of evaluation functions, $f_1, ..., f_\mathrm{m}$. For each individual assessment at a given scenario, the lower rating means a better evaluation. The formulation of the multi-criteria optimization problem is expressed in the decision space. This is a way for representating the decision problem, where the goal is to choose the right decision, given a set of criteria associated with decision's risk.

There is a transformation $f = (f_1, f_2, ..., f_m)$ of a set of feasible decisions, X_0, into a set of achievable assessment vectors, Y_0. The problem of choosing the best decision arises naturally. The choice of decision only considers grade vectors and decisions with identical grade vectors that are equally good. Thus, the problem of determining the best decision can be limited to the issue of choosing the best grade vector in the set of achievable grades (achievable grade vectors):

$$Y_0 = \{y : \; y = f(x), \; x \in X_0\}$$

This leads to a multi-criteria model in the assessment space:

$$\min_x \{y = (y_1, ..., \mathrm{y_m}) : \; y_i = f_i(x) \; \forall i, \; x \in X_0\} \tag{2}$$

where grades are directly specified as individual variables.

Each vector x in the set X_0 corresponds to the vector y for the set Y_0. The vector from the set Y_0 is selected and one sees the decision from the set X_0.

3. Symmetrically efficient solutions

This chapter extended the way of defining a symmetrically effective decision compared to the work [1]. This chapter provides the basic definition of a symmetrically effective decision. It is a decision that is a solution to a specific multi-criteria optimization problem, a problem used to support decisions under risk. Decision assessments must meet an additional condition—the condition of anonymity of preference relationships.

The model of the decision problem under risk in the form of a multi-criteria optimization problem imposes additional properties of preference relations and, consequently, limits the choice of decisions to an appropriate subset of the entire set of effective decisions. In the problem of making decisions under risk, minimizing all assessments is equally important.

Decision problems under risk are, when the decision is based on minimization of a vector outcome with various realizations under several scenarios. The preference model leads to the Pareto efficiency with respect to the realizations under scenarios understood as multiple criteria. The case of equally probable scenarios leads to the concept of symmetric optimization (efficiency) of multi-criteria corresponding to realizations under scenarios. The solution should have the feature of anonymity: no distinction is made between results that differ in their orientation coordinates. This solution of the problem, called a symmetrically efficient decision, is an efficient decision that possesses an additional property, that is, that of preference relation anonymity.

Nondominated solutions (*optimum Pareto*) are defined with the use of preference relations which answer the question of which one of a given pair of evaluation vectors $y^1, y^2 \in R^m$ is better. This is the following relation:

$$y^1 \succ y^2 \Leftrightarrow y_i^1 \leq_i^2 \forall i = 1, \dots, m \wedge \exists j \; y_j^1 < y_j^2 \tag{3}$$

The vector of evaluation $\hat{y} \in Y_0$ is called the *nondominated vector*, provided there is no vector $y \in Y_0$ such that $\hat{y}$ is dominated by y. The domination structure is shown in **Figure 1**.

The set of *nondominated solutions* is defined as follows:

$$\hat{Y}_0 = \{\hat{y} \in Y_0 : (\hat{y} + \tilde{D}) \cap Y_0 = \varnothing)\} \tag{4}$$

where $\tilde{D}$ is a positive cone without the top. The positive cone can be $\tilde{D} = R_+^m$.

The set $\hat{Y}_0$ is shown in **Figure 2**.

A decision $\hat{x} \in X_0$ is called *an efficient decision* (*Pareto optimal*) if there is no $\overline{x}$ such that $y_i = f_i(\overline{x}) \leq \hat{y}_i = f_i(\hat{x})$ for $i = 1, \dots, m$ with strict inequality for at least one i [5, 8].

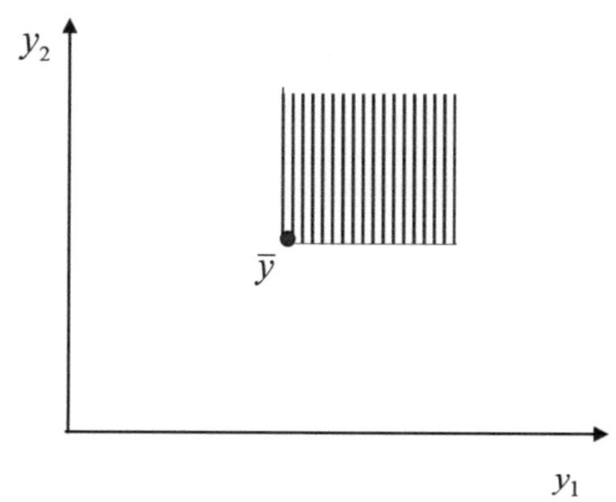

Figure 1.
Dominance structure in R^2.

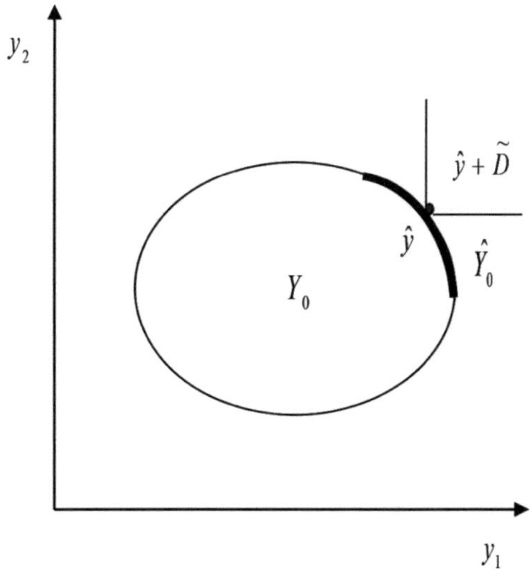

Figure 2.
Nondominated solutions.

In the problem with homogeneous and equally important assessments, the relation of the decision maker preferences should be impartial due to individual assessment functions. That is, for a given set of evaluation functions, only the distribution of the values achieved by these functions for a given decision is important, and it is not important which function it took. This requirement is formulated mathematically as a property of the anonymity of preference relationships. The risk assessment vector should meet the property of anonymity.

The relation is called an *anonymous* (*symmetric*) relation if, for every vector $y = (y_1, y_2, \dots, y_m) \in R^m$ and for any permutation, P, of the set $\{1, \dots, m\}$, the following is true:

$$\left(y_{P(1)}, y_{P(2)}, \dots, y_{P(m)}\right) \approx (y_1, y_2, \dots, y_m) \tag{5}$$

We look at the whole with the help of an anonymous preference relationship, and all scenarios are considered rather than on individual results for given scenarios. Anonymous preference relationship is a superstructure over a preference relationship—an additional condition of anonymity is added.

A nondominated vector satisfying the anonymity property is called *a symmetrically nondominated vector*. The set of symmetrically nondominated solutions is marked as follows: $\hat{Y}_{os}$. In the decision space, symmetrically efficient decisions are specified. The decision $\hat{x} \in X_0$ is called *a symmetrically efficient decision*, if the corresponding evaluation vector $\hat{y} = f(\hat{x})$ is a symmetrically nondominated vector. The set of symmetrically efficient decisions is marked as follows: $\hat{X}_{os}$ [9, 10].

The domination structure of symmetric dominance depends on the location of an evaluation vector, y, relative to the line $y_1 = y_2 = \dots = y_m$. The domination structure is shown in **Figure 3**.

The relation of symmetric domination can be expressed as the domination of evaluation vectors with coordinates ordered in no decreasing order. This can be formalized with the map $T: R^m \rightarrow R^m$ such that $T(y) = (T_1(y), T_2(y), \dots, T_m(y))$, where $T_1(y), \geq T_2(y) \geq, \dots, \geq T_m(y)$ and a permutation, P, of the set $\{1, \dots, m\}$ exists such that $T_i(y) = y_{P(i)}$ for $i = 1, \dots, m$.

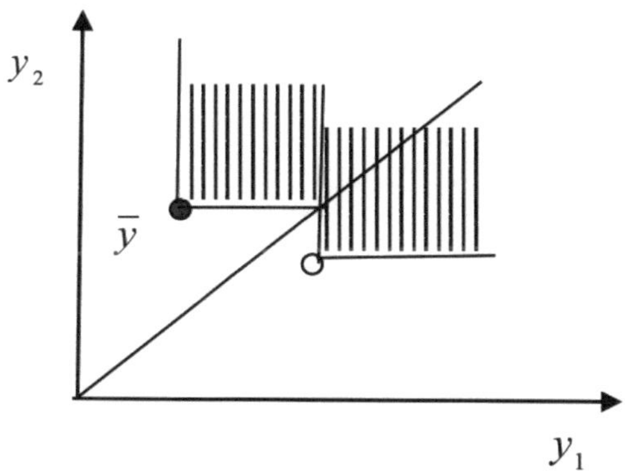

Figure 3.
Symmetric dominance structure in R^2.

The evaluation vector y' symmetrically dominates the vector y'' if the following condition is satisfied:

$$y^1 \succ_a y^2 \Leftrightarrow T(y^1) \leq T(y^2) \tag{6}$$

The relation of symmetrical domination $\succ_a$ is a simple vector domination for evaluation vectors with no decreasing coordinates of evaluation vector [9, 10].

For the problem of decisions under risk expressed in the form of a multi-criteria optimization problem, the solution is a set of symmetrically effective decisions.

4. Technique of generating symmetrically efficient decisions

This chapter extended the way of defining a symmetrically effective decision compared to the work [1]. This chapter discusses how to support decisions under risk. It is an interactive IT system that processes relevant data for a given decision situation and assists the decision maker in recognizing the decision problem in the sense of understanding his own preferences. The decision maker's role is paramount. The system is not a substitute for the decision maker at any stage of decision-making. Such a system is to support, not replace in the final selection of the decision maker.

In multi-criteria decision problems, the relation of preferences is not known a priori, and therefore the final choice of solution can be made only by the decision maker. Given the numerous set of solutions, this selection is made using the appropriate interactive information system—the decision support system. Such a system processes important data for a given decision situation but also supports the decision maker in recognizing the decision problem in the sense of understanding his own preferences. The decision maker's role is paramount. The system is not a substitute for the decision maker at any stage of decision-making. Decision support system is to support, not replace in the final selection of the decision maker. In the problem of multi-criteria optimization, you cannot impose an optimal solution on the decision maker, you should support it—give the decision maker the opportunity to review such solutions that give the best results—symmetrically nondominated solutions. The decision maker chooses the decision by looking at the symmetrically nondominated set. This system enables a controlled review of the set of symmetrically efficient solutions. On the basis of the values of certain control parameters given by the decision maker, the system presents various solutions that are symmetrically efficient for analysis.

Solutions of a symmetrically efficient multi-criteria problem can be determined by solving the optimization of a multi-criteria problem:

$$\min\left\{\left(y_1,\ldots,y_m\right):x\in X_0\right\} \tag{7}$$

with the scalarizing function $y: R^m \rightarrow R$ defining the preference relation:

$$y^1 \succ_a y^2 \Leftrightarrow s\left(y^1\right) > s\left(y^2\right) \tag{8}$$

If the relation fulfills the condition of anonymity, the efficient solution generated by this scalarization is also a symmetrically efficient solution to the multi-criteria problem (1).

Symmetrically efficient decisions for a multiple criteria problem (1) are obtained by solving a special problem in multi-criteria optimization, that is, a problem with coordinates of the vector of evaluation arranged in a no decreasing order. This problem is as follows:

$$\min_y \left\{\left(T_1(y), T_2(y), \ldots, T_m(y)\right) : y\in Y_0\right\} \tag{9}$$

where $y = \left(y_1, y_2, \ldots, y_k\right)$ is an evaluation vector,

$T(y) = \left(T_1(y), T_2(y), \ldots, T_m(y)\right)$, where $T_1(y), \geq T_2(y) \geq, \ldots, \geq T_m(y)$ is an ordered evaluation vector,

Y_0 is the set of evaluation vectors.

An efficient solution of multi-criteria optimization problem (9) is a symmetrically efficient solution of the multi-criteria problem (1).

The method of determining individual symmetrically efficient decisions involves the solution of a parametric scalarization of a multi-criterion problem. This is a problem of single objective optimization using a specially created scalarizing function of two variables: the evaluation vector, $y\in Y$, and control parameter, $\overline{y}\in\Omega\subset R^m$; thus, we have $s: Y_0 \times \Omega \rightarrow R^1$:

$$\min_x \left\{s\left(y_1,\ldots,y_m\right):x\in X_0\right\} \tag{10}$$

The parameter $\overline{y} = \left(\overline{y}_1, \overline{y}_2, \ldots, \overline{y}_m\right)$ is available to the decision maker, enabling him or her to review the set of symmetrically efficient solutions.

To ensure the anonymity of the relationship, it is necessary and sufficient that the scalarizing function is symmetrical, that is,

$$s\left(y_{P(1)}, y_{P(2)}, \ldots, y_{P(m)}\right) \approx s\left(y_1, y_2, \ldots, y_m\right) \tag{11}$$

for any permutation, P, of the set $\{1, \ldots, m\}$.

Complete and sufficient parameterization of the set of symmetrically efficient solutions can be achieved, using the method of the reference point for problem (9). In this method, aspiration levels are applied as control parameters. An aspiration level is a value of the evaluation function that satisfies the decision maker.

The scalarizing function defined in the method of the reference point is as follows:

$$s(y,\overline{y}) = \min_{1\leq i\leq m}\left(T_i(y) - T_i(\overline{y})_i\right) + \varepsilon\cdot\sum_{i=1}^{m}\left(T_i(y) - T_i(\overline{y})_i\right) \tag{12}$$

where $y = \left(y_1, y_2, \ldots, y_m\right)$ is an evaluation vector,

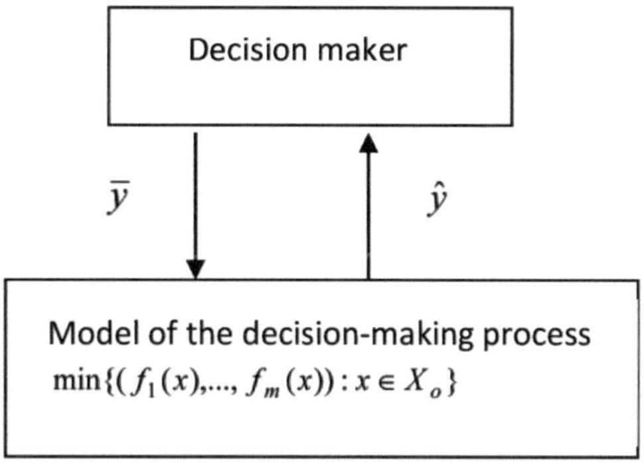

Figure 4.
The method of supporting decision selection.

$T(y) = (T_1(y), T_2(y), ..., T_m(y))$, where $T_1(y), \geq T_2(y) \geq, ..., \geq T_m(y)$ is a no decreasing ordered evaluation vector,

$\bar{y} = (\bar{y}_1, \bar{y}_2, ..., \bar{y}_k)$ is a vector of aspiration levels,

$T(\bar{y}) = (T_1(\bar{y}), T_2(\bar{y}), ..., T_m(\bar{y}))$, where $T_1(\bar{y}) \geq T_2(\bar{y}) \geq, ..., \geq T_m(\bar{y})$ is a no decreasing order vector levels of aspiration,

ε is an arbitrary small, positive adjustment parameter.

This kind of scalarizing function is called *a function of achievement*. The aim is to find the solution closest to the specific requirements, that is, the aspiration levels. Maximizing this function determines the symmetrically efficient solution, $\hat{y}$, and the symmetrically efficient decision, $\hat{x}$. Note that the symmetrically efficient solution, $\hat{x}$, depends on the aspiration level, $\bar{y}$ [9, 11].

The solution to the multi-criteria optimization problem is a set of efficient solutions. The choice of solution should be made by the decision maker using an IT system. Such a system allows him to browse the entire set of solutions and make choices freely. The final choice of the solution among the set of efficient solutions can only take place based on the user's preferences. A tool for searching the set of solutions is the function (12). The maximum of this function depends on the parameter, $\bar{y}$, which is used by the decision maker to select a solution. The method of supporting decision selection is an iterative method consisting of the alternating performance of:

- calculations, that is, finding another symmetrically efficient solutions;
- interaction with the system, that is, dialog with the decision maker, which is a source of additional information about his or her preferences.

The method of supporting decision selection is shown in **Figure 4**.

This method of supporting decision-making, which does not impose a rigid scenario for the analysis of the decision-making problem upon the decision maker, enables modification of his or her preferences during the analysis of the problem. The decision maker plays a key role in the decision-making process.

5. Example: selecting a decision

The problem of selecting a decision is shown in order to illustrate the method of supporting a decision under risk [12]. The costs of 10 alternatives in three scenarios

Decision	S_1	S_2	S_3
Decision x_1	59	65	75
Decision x_2	50	58	71
Decision x_3	68	72	60
Decision x_4	69	72	62
Decision x_5	53	60	63
Decision x_6	51	59	65
Decision x_7	68	71	77
Decision x_8	56	57	75
Decision x_9	62	58	80
Decision x_{10}	62	55	70

Table 1.
Scenarios of 10 decisions.

are presented in **Table 1**. The probabilities of particular scenarios are as follows: P1 = 0.3, P2 = 0.6, and P3 = 0.1.

The decision maker's problem is to select one of 10 decisions with three possible future scenarios. Since the configuration of conditions that will apply during the decision is unknown, this problem is a selection decision under risk [13].

The problem of decision-making under risk is modeled as a multi-criteria optimization problem:

$$\min_x x\{y^1,y^2,y^3,y^4,y^5,y^6,y^7,y^8,y^9,y^{10} : x \in \{x_1,x_2,x_3,x_4,x_5,x_6,x_7,x_8,x_9,x_{10}\}\}, \tag{13}$$

where the results of particular decisions are the following vectors:

$$\begin{aligned}
y^1 &= (59,65,75) \text{ for decision } x_1,\\
y^2 &= (50,58,71) \text{ for decision } x_2,\\
y^3 &= (68,72,60) \text{ for decision } x_3,\\
y^4 &= (69,72,62) \text{ for decision } x_4,\\
y^5 &= (53,60,63) \text{ for decision } x_5,\\
y^6 &= (51,59,65) \text{ for decision } x_6,\\
y^7 &= (68,71,77) \text{ for decision } x_7,\\
y^8 &= (56,57,75) \text{ for decision } x_8,\\
y^9 &= (62,58,80) \text{ for decision } x_9,\\
y^{10} &= (62,55,70) \text{ for decision } x_{10},
\end{aligned}$$

in which particular coordinates of evaluation vectors occur with probabilities: $P_1 = 0.3$, $P_2 = 0.6$, and $P_3 = 0.1$.

The problem consists in selecting a decision for which the evaluation vector has the minimum value in the sense of symmetrical dominance.

The repeating of relevant scenarios results in a situation in which the probability of each scenario is the same and, that is, $P = 1/10$. The result is a problem equivalent to the starting problem in which the results for each decision, namely: $x_1,x_2,x_3,x_4,x_5,x_6,x_7,x_8,x_9,x_{10}$ are the following evaluation vectors with equally probable coordinates:

$$y^1 = (59, 59, 59, 65, 65, 65, 65, 65, 65, 75),$$
$$y^2 = (50, 50, 50, 58, 58, 58, 58, 58, 58, 71),$$
$$y^3 = (68, 69, 68, 72, 72, 72, 72, 72, 72, 60),$$
$$y^4 = (69, 69, 69, 72, 72, 72, 72, 72, 72, 62),$$
$$y^5 = (53, 53, 53, 60, 60, 60, 60, 60, 60, 63),$$
$$y^6 = (51, 51, 51, 59, 59, 59, 59, 59, 59, 65),$$
$$y^7 = (68, 68, 68, 71, 71, 71, 71, 71, 71, 77),$$
$$y^8 = (56, 56, 56, 57, 57, 57, 57, 57, 57, 75),$$
$$y^9 = (62, 62, 62, 58, 58, 58, 58, 58, 58, 80),$$
$$y^{10} = (62, 62, 62, 55, 55, 55, 55, 55, 55, 70).$$

In order to compare the vectors in the sense of symmetrical dominance, the coordinates of vectors are ordered in no decreasing order and the results are the following evaluation vectors for each decision:

$$T(y^1) = (75,\ \ 65, 65, 65, 65, 65, 65, 59, 59, 59),$$
$$T(y^2) = (71, 58, 58, 58, 58, 58, 58, 50, 50, 50),$$
$$T(y^3) = (72, 72, 72, 72, 72, 72, 68, 69, 68, 60),$$
$$T(y^4) = (72, 72, 72, 72, 72, 72, 69, 69, 69, 62),$$
$$T(y^5) = (63, 60, 60, 60, 60, 60, 60, 53, 53, 53),$$
$$T(y^6) = (65, 59, 59, 59, 59, 59, 59, , 51, 51, 51),$$
$$T(y^7) = (77,\ \ 71, 71, 71, 71, 71, 71, , 68, 68, 68),$$
$$T(y^8) = (75, 57, 57, 57, 57, 57, 57, , 56, 56, 56),$$
$$T(y^9) = (80, 62, 62, 62, 58, 58, 58, 58, 58, 58),$$
$$T(y^{10}) = (70, 62, 62, 62, 55, 55, 55, 55, 55, 55).$$

The set of symmetrically nondominated vectors is as follows: $\hat{Y}_{os} = \{y^2, y^5, y^6, y^8, y^{10}\}$. Five decisions x_2, x_5, x_6, x_8, and x_{10} are symmetrically efficient decisions. When making a selection, one should choose from among them and the decisions x_1, x_3, x_4, x_7, and x_9 should be rejected regardless of individual preferences. These five decisions are incommensurate with respect to a symmetrical preference relation. The choice between them depends on the individual preferences of the decision maker.

The method of the reference point for the problem with coordinates of the evaluation vector arranged in no decreasing order is used to determine the solution of the problem (13). The decision maker controls the selection of an investment project through the levels of aspiration by specifying the desired values of the aspiration vector for each scenario: $\bar{y} = (\bar{y}_1, \bar{y}_2, \bar{y}_3)$, where $\bar{y}_1$ is a level of the aspiration value for scenario 1, $\bar{y}_2$ is a level of the aspiration value for scenario 2, and $\bar{y}_3$ is a level of the aspiration value for scenario 3.

Iteration	
1. Aspiration level $\bar{y}$ Solution $\hat{x}$	$\bar{y} = (50, 55, 60)$ Decision x_6
2. Aspiration level $\bar{y}$ Solution $\hat{x}$	$\bar{y} = (55, 60, 60)$ Decision x_5
3. Aspiration level $\bar{y}$ Solution $\hat{x}$	$\bar{y} = (55, 56, 68)$ Decision x_2
4. Aspiration level $\bar{y}$ Solution $\hat{x}$	$\bar{y} = (55, 56, 74)$ Decision x_8
5. Aspiration level $\bar{y}$ Solution $\hat{x}$	$\bar{y} = (62, 56, 68)$ Decision x_{10}
Source: own calculations.	

Table 2.
Interactive analysis of the search for a decision.

The multiple-criteria analysis is presented in **Table 2**.

At the beginning of the selection, the decision maker identifies the aspiration levels as the best values that can be achieved separately for each scenario, and in subsequent iterations, he or she changes the aspiration levels depending on his or her preferences.

In the first iteration, the decision maker determines the preferences as an aspiration level equal to the vector $\bar{y} = (50, 55, 60)$ and obtains the decision solution, x_6, as the solution. In the second iteration, the decision maker reduces the requirements for scenarios 1 and 2 without changing the requirement for scenario 3, states the vector $\bar{y} = (55, 60, 60)$ as the aspiration level, and obtains decision x_5 as the solution. In the third iteration, the decision maker does not change the requirements for scenario 1, increases them for scenario 2, and reduces them for scenario 3, and he or she states the vector $\bar{y} = (55, 56, 68)$ as the aspiration level and obtains decision x_2 as the solution. In the fourth iteration, the decision maker does not change the requirements for scenarios 1 and 2 and reduces the requirement for scenario 3. He or she states the vector $\bar{y} = (55, 56, 74)$ as the aspiration level and obtains decision x_8 as the solution. In the fifth iteration, the decision maker reduces the requirements for scenario 1 leaves the requirements for scenario 2 unchanged, and increases the requirements for scenario 3. He or she states the vector $\bar{y} = (62, 56, 68)$ as the aspiration level and obtains decision x_{10} as the solution.

The final selection of a specific solution depends on the decision maker's preferences. The example given here shows that the method enables the decision maker to discover his or her decision-making capabilities in the course of interactive analysis and obtain a satisfactory solution.

6. Conclusions

In the decision-making process, risk plays a significant role, influencing the final result of the decision. The decision maker should be able to analyze them when making decisions. Using his experience and information from the environment, he should make such decisions that will not bring unnecessary threat (risk) to the effects of the decision. Despite the use of objectified tools optimizing decision-making processes in the choice of solution, ultimately the decision maker takes responsibility for the decisions taken.

The chapter presents a method for the decision made under risky situations. The risk is introduced to the model with a set of scenarios with specified probabilities. The choice is made by solving the problem of multi-criteria optimization. This provides a systematic procedure to help a decision maker choose the most desirable and satisfactory decision under risk situations. Therefore, using this way, a decision can be made according to the decision maker's preference. This method is characterized by:

- The use of reference point method, that is, the concepts of aspiration levels and minimization of the achievement function to organize interaction with the decision maker.

- The assumption that the decision maker's preferences are not fully formed changes during the decision-making process, while the main problem of the decision support system is to support the decision maker's learning rather than the final act of choice.

- The method gives a whole set of solutions symmetrically effective decisions and allows the decision maker a free choice. This procedure does not replace the decision maker in making decisions. The whole decision-making process is controlled by the decision maker.

Author details

Andrzej Łodziński
Warsaw University of Life Sciences, Warsaw, Poland

*Address all correspondence to: andrzej_lodzinski@sggw.edu.pl

References

[1] Łodziński A. The method of supporting decisions under risk based on multi-objective optimization. In: Quantitative Methods in Economics. Vol. 14. Issue. 2. Warsaw: Warsaw University of Life Sciences – SGGW. 2013. pp. 39-50

[2] Abdellaoui M, Hey JD, editors. Advances in Decision Making under Risk and Uncertainty. Berlin Heidelberg: Springer-Verlag; 2008

[3] Cox JC, Sadiraj V. On the coefficient of variation as a criterion for decision under risk. Journal of Mathematical Psychology. 2010;**54**:387-394

[4] Crouhy H, Galai D, Mark R. Risk Management. New York: McGraw-Hill; 2001

[5] Goodwin P, Wright G. Decision Analysis (in Polish). Warsaw: Oficyna a Wolters Kluwer Business; 2001

[6] Keeney L, Raiffa H. Decisions with Multiple Objectives. Preferences and Value Tradeoffs. Cambridge University Press; 1993.

[7] Krawczyk T. Modeling of Investment Risk (in Polish). Warsaw: CeDeWu; 2013

[8] Wierzbicki AP. A mathematical basis for satisficing decision making. Mathematical Modelling. 1982;**3**: 391-405

[9] Mun J. Modelling Risk. New York: Wiley & Sons, Inc; 2014

[10] Trzaskalik T. Multi-Criteria Decision Support: The Methods and Applications (in Polish). Warsaw: PWE; 2014

[11] Kunreuther HC, Linnerooth J, Lathrop J, Hermann Atz H, Macgill S, Mandl C, et al.. with a Foreword by Douglas M.Risk Analysis and Decision Processes. Berlin/Heidelberg/New York/ Tokyo: Springer-Verlag; 1983

[12] Matos MA. Decision under risk as a multi-criteria problem. European Journal of Operational Research. 2007; **181**:1516-1529

[13] Wierzbicki A, Makowski N, Wessels J. Model-Based Decision Support Methodology with Environmental Applications. Laxenburg/Dordrecht: IIASA Kluwer; 2000

Chapter 10

Collaborative Decision-Making Processes for Cultural Heritage Enhancement: The Play ReCH Platform

Maria Cerreta, Gaia Daldanise, Paolo Di Lauro and Ludovica La Rocca

Abstract

These days, cultural heritage is one of the topics at the center of the urban sustainability agenda. Current economic and urbanization trends place significant pressure on urban resources, systems, and infrastructures and demand for novel approaches in governing, financing, and monitoring urban performances with particular attention to abandoned, unused, or underutilized cultural heritage, defined "waste heritage." In this perspective, cities are laboratories where innovative and collaborative approaches can be tested, and culture-led processes can be implemented consistent with circular economy principles. In order to structure and activating collaborative decision-making processes for regeneration and adaptive transformation of cultural heritage, gamification assumes a central role. The chapter analyzes the interaction among gamification and collaborative decision-making processes relevant to support the enhancement of cultural heritage and describes the Play ReCH (Reuse Cultural Heritage) platform, winner of the 2019 Welfare Che Impresa call, activated with the purpose to promote a cultural creative enterprise and include cooperation and innovation in cultural heritage regeneration processes. Play ReCH allows rethinking the management model of cultural heritage reuse through gamification processes in combining technology and reality, involving city users within creative processes.

Keywords: cultural and creative enterprises, intellectual property, co-evaluation, gamification, cultural heritage

1. Introduction

In 2016, the 2030 Agenda for Sustainable Development [1], at the 11th point, defines cities as centers for new ideas, commerce, culture, science, productivity, and social development, inviting to strengthen efforts to protect and safeguard cultural and natural heritage in order to build sustainable cities and communities. In Europe, some disadvantaged areas, despite criticism, offer fertile ground for developing new cultural approaches for learning and sharing instruments to face some relevant conflicts related to unemployment, industrial abandoned areas, deprived historical centers, different interests of the public and private, and inclusion of foreign citizens. To counter these kinds of conflicts, the creative cultural enterprise [2, 3], with its multidisciplinary nature is considered a key actor for attracting and implementing a new workforce and innovative forms of experimentation on multicultural identities, knowledge economies, and innovative territorial services. Indeed, creative capital could be able to optimize local cultural resources for rebuilding relationships among communities, values, and places [4–7] in a productive way, thereby enhancing culture-led urban regeneration processes locally embedded [8, 9]. Within this processes, especially the reuse of cultural heritage [10] is strictly linked to new forms of welfare that consider the proximity of services, activities, and places [11] such as a new model of social, territorial development.

In this perspective, cities become laboratories where to test innovative and culture-led approaches, in which cultural factors are catalysts in the production and regeneration processes, and where particular attention can be dedicated to "waste heritage," considered as abandoned, unused, or underutilized heritage that needs new approaches and models for use and management. Indeed, culture as an engine for urban economic growth has become part of the new orthodoxy with which cities try to improve their competitive position by aiming for greater attractiveness in close correlation with their identity. A sustainable, regenerative development model for cultural heritage can, therefore, be achieved by introducing culture as a strategic investment sector. At the same time, there is a need to activate and promote processes capable of enhancing cultural heritage by making use of approaches and tools that include innovation and cooperation as essential components. In this context, collaborative decision-making processes that consider multi-methodological approaches and are based on the methods proper to gamification represent an interesting opportunity to make these components operational. New decision-making contexts are emerging, capable of generating different forms of value, which include not only economic values but also intrinsic, social, and shared values. Identify and assess "complex shared social values" [12, 13] also mean to explore the cultural demand and the creative production, in order to adopt new decision-making tools and new economic actors, recognizing the crucial role of cultural and creative enterprises.

In this perspective, our key research questions are how building a creative cultural enterprise for cultural heritage enhancement able to implement innovative evaluation and management models in terms of business value proposition [14]? How can gamification improve collaborative decision-making processes for the enhancement of cultural heritage?

Taking into account our research questions, the contribution was structured as follows: an introduction to cultural demand and creative production, where evaluation and monitoring frameworks are analyzed (Section 1.2); a presentation of legal framework for the intellectual property strictly linked to cultural and creative enterprises and their original creative content (Section 1.2); the description of the methodological approach based on collaborative decision-making processes and gamification for cultural heritage enhancement (Section 2), considering two sections (Sections 2.1 and 2.2), respectively, related to complex social shared values in a co-evaluation approach and a game-thinking-centered approach for cultural creative enterprises; the results about the articulation of Play ReCH platform experiment conceived as innovative model of cultural creative enterprise, where the methodological approach proposal was tested (Section 3), articulated in two sections, the first that concerns the description of the Play ReCH platform, and the second, the activation of a new model of cultural creative enterprise for cultural heritage enhancement; and the last section is dedicated to the discussion and conclusions on the whole process (Section 4).

1.1 Cultural demand and creative production: evaluation and monitoring frameworks

In the last decades, the "demand of culture" has grown, and the cultural heritage is seen as a territorial system of complex values [5], goods, services, and relationships, which become part of the local identity, through art, history and landscape, and immaterial culture as food and wine, crafts, and traditions [15].

In different contexts, there is a need to build new common identities [16, 17], generating a closer relationship between the third sector, public administration, private actors, and citizens toward the definition and implementation of innovative urban policies [18], in which cultural heritage plays a key role [10, 13].

Indeed, the new European economic strategy, from the "Lisbon Strategy" to the "Horizon 2020" programs and their future developments, aims to achieve three main shared objectives: smart growth, sustainable growth, and inclusive growth. Within these three objectives, culture becomes a strategic priority [19], and it is also recognized the key role of cultural and creative enterprises [20–24] able to generate new knowledge and culture through creativity, skills, and talent, building new forms of wealth and jobs and transforming some critical issues into opportunities. Such forms of enterprise produce goods, services, and assets that are recognized not only for their economic value [25] but also and above all because of their intrinsic value [26], and for their capacity to activate value chains [27]. Most cultural enterprises are supported by public funds and produce cultural goods for use (e.g., museums, archives, etc.) [28], while creative enterprises are supported by the market and produce goods (e.g., design products, architecture, and fashion) for consumption. In Italy, the definition of cultural and creative enterprises [23] is particularly related to the historical-architectural and artistic heritage, to entertainment, music, and contemporary arts. At the same time, information and communications are taken into account, integrating ICT in the production of cultural services and goods.

The common characteristic of the different definitions and models of enterprise is the ability to create shared value [29] functional to a company's competitive position. Shared value optimizes and uses specific resources and skills to build economic value through the creation of social value, not only generating employment opportunities but also building process and product innovation [11] through new models of shared responsibility: from Corporate Social Responsibility and Community Social Responsibility to Territorial Social Responsibility [30] for building complex shared social values [12, 13, 26].

The ability to build complex shared social values demonstrates the role of culture in implementing sustainable development as also analyzed by UNESCO Global Report on Culture for Sustainable Urban Development [31].

At the global level already in 2009, UNESCO developed "The 2009 UNESCO Framework for Cultural Statistics (FCS)" as a tool to measure the social and economic dimensions of the cultural phenomenon through a "culture cycle model" [32] of creation, production, dissemination, transmission, and consumption. This framework defines "culture" as a set of distinctive spiritual, material, emotional, intellectual, and emotional characteristics of a social group or society that include value systems, ways of life, traditions, and beliefs [32].

Subsequently, UNESCO also developed the "Culture for Development Indicators (CDIS)" project, which proposes a new methodology to demonstrate the role of culture as a driver of sustainable development processes, based on empirical data [33]. This project is based on the "Convention on the Protection and Promotion of the Diversity of Cultural Expressions" [25], which implements the integration of culture and the promotion of the diversity of cultural expressions in development policies at all levels. The CDIS project addresses culture in terms of values and norms that guide human action and not only as a productive or leisure sector. The CDIS project aim is assessing the multidimensional role of culture in sustainable development, so it encourages an inclusive view of culture's interactions in development by exploring not only economic benefits but also intangible benefits such as inclusion, tolerance, and social cohesion. In particular, the seven key policy dimensions exanimated by the CDIS methodology are (1) Economy, (2) Education, (3) Governance, (4) Social participation, (5) Gender equality, (6) Communication, and (7) Heritage. Some examples of core indicators are presented in **Table 1** and are useful to understand the different ways for describing and assessing cultural activities and processes.

At the same time, at the European level, the ESSnet-Culture framework [34] is being developed to help EU countries in building their specific cultural framework by identifying the different areas covered by cultural statistics. Each country adopts UNESCO FCS definitions of cultural domains, and this allows international comparability of countries' data (**Figure 1**).

The ESSnet update for European cultural statistics—previously defined in 2000 by the LEG-Culture framework—is the first step toward a common framework useful for the production of comparable European data on different culture-related topics [35]. The ESSnet-Culture structure is based on three key concepts: cultural domain, function, and dimension.

There are 10 cultural domains, and they consist in a set of cultural activities, practices, and products focused on artistic expressions: Heritage; Archives; Libraries; Book and Press; Visuals Arts; Performing Arts; Audiovisual and Multimedia; Architecture; Advertising; and Art crafts. The six functions identified concern a mapping of the main cultural activities identifiable with existing statistical and economic classifications: creation, as the elaboration of original cultural content and ideas; production/publishing, as a part of the economic cycle of the creative idea inserted in the production as original content (cultural product or service) that becomes reproducible also for other users through advertising; dissemination/trade, which makes the product/service available online and offline to consumers; preservation, a phase that includes both activities to protect and restore cultural heritage and digitalization; education, as the transfer of skills and abilities within cultural activities; and management/regulation, intended as a set of activities able to create an enabling environment for operators, spaces, and cultural services [34]. Finally, the third key concept for the ESSnet framework concerns dimensions, intended as approaches closely linked to cultural activities, such as social practices and participation, employment, and consumption.

A further crucial definition introduced by this European framework is the concept of cultural activity as an activity based on cultural values and/or artistic expressions, including both market-oriented and noncommercial activities. Such activities may be carried out by individuals, companies, organizations, groups, or professionals within a specific cultural field and according to the function necessary for its realization. Furthermore, within the definition, the framework introduces statistical classifications, mainly economic NACE Rev.2 [36] for the production of data and the measurement of these activities through NACE codes representing the cultural production sector. At the national accounting level, there is also a differentiation between market and nonmarket sectors.

Another European study that has become crucial for the connection of cultural sectors to urban development is the "Cultural and Creative Cities Monitor" [37], which aims to assess and monitor the performance of European "cultural and creative cities" in terms of jobs and economic growth. The tool is based on 29 indicators organized in nine dimensions, reflecting three key dimensions of cultural and creative cities (Cultural Vibrancy, Creative Economy, and Enabling

Key policy dimensions	Indicators	Unit of measure
1. Economy	1.1 Contribution of cultural activities to GDP	Percentage of the contribution of private and formal cultural activities to Gross Domestic Product
	1.2 Cultural employment	Percentage of persons engaged in cultural occupations within the total employed population
	1.3 Household expenditures on culture	Percentage of household final consumption expenditures on cultural activities, goods, and services
2. Education	2.1 Inclusive education	Index of average years of schooling of the population between the ages of 17 and 22, adjusted to reflect inequalities
	2.2 Multilingual education	Percentage of instructional hours dedicated to promoting multilingualism in relation to the total number of instructional hours dedicated to languages (Grades 7–8)
	2.3 Arts education	Percentage of instructional hours dedicated to arts education in relation to the total number of instructional hours (Grades 7–8)
	2.4 Professional training in the culture sector	Index of coherency and coverage of technical and vocational education and training (TVET) and tertiary education in the field of culture
3. Governance	3.1 Standard-setting framework for culture	Index of development of the standard-setting framework for the protection and promotion of culture, cultural rights, and cultural diversity
	3.2 Policy and institutional framework for culture	Index of development of the policy and institutional framework for the protection and promotion of culture, cultural rights, and cultural diversity
	3.3 Distribution of cultural infrastructures	Distribution of selected cultural infrastructures relative to the distribution of the country's population in administrative divisions immediately below state level
	3.4 Civil society participation in cultural governance	Index of the promotion of the participation of cultural professionals and minorities in the formulation and implementation of cultural policies, measures, and programs that concern them
4. Social participation	4.1 Participation in going-out cultural activities	Percentage of the population who have participated at least once in a going out cultural activity in the last 12 months
	4.2 Participation in identity-building cultural activities	Percentage of the population who have participated at least once in an identity-building cultural activity in the last 12 months
	4.3 Tolerance of other cultures	Degree of tolerance within society toward people from different cultural backgrounds
	4.4 Interpersonal trust	Degree of interpersonal trust
	4.5 Freedom of self-determination	Median score of perceived freedom of self-determination

Table 1.
Culture for development indicators (CDIS).

Environment) using comparable quantitative and qualitative data [38]. The 2017 edition covered the monitoring of 168 cities in 30 European countries.

At the same time, other studies are oriented to build a set of ad hoc indicators to measure the creativity of EU Member States such as the KEA European Affairs European Creativity Index (ECI) [39], which takes into account a number of factors

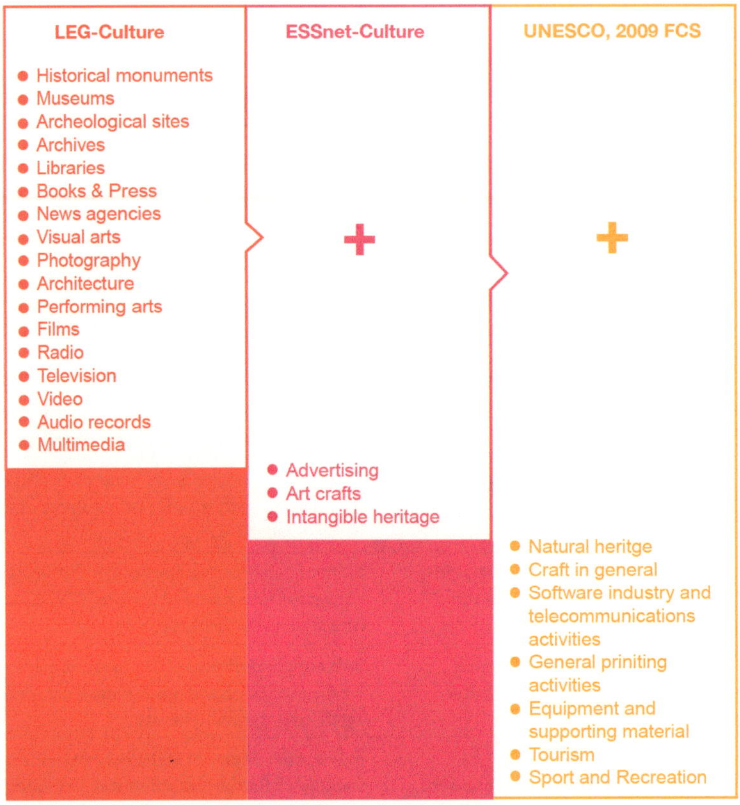

Figure 1.
Cultural domains covered by European and UNESCO frameworks for cultural statistics (source: [34]).

on the cultural dimension, often not included in the other indices, such as the study of art subjects in schools, cultural supply, participation in cultural events, financial support for creativity, and the role of technology. ECI is based on 32 indicators structured in six pillars: Pillar I—Human Capital; Pillar II—Openness and Diversity; Pillar III—Social Environment; Pillar IV—Technology; Pillar V—Institutional Environment; and Pillar VI—Products of Creativity (**Figure 2**).

In Italy, the main data sources on culture, creativity, participation, and cultural employment concern the following surveys:

- ISTAT survey *Aspects of everyday life* (annual frequency since 1993);
- ISTAT survey *Citizens and leisure* (regular frequency: 1995, 2000, 2006, and 2015);
- ISTAT *Culture and leisure survey* (regular frequency: 2017, 2018, and 2019); and
- Symbola Foundation and Unioncamere survey *I am culture—the Italy of quality and beauty defies the crisis* (annual frequency: 2017, 2018, and 2019).

In 2018, compared to 2017, cultural participation increased from 64.1 to 64.9%, with growth concentrated among those who claim to have participated in more than four events in the last 12 months (from 23.2 to 24.5%). In particular, the increase mainly concerns visitors in archeological sites and monuments and who have attended music concerts (not classical music) [40].

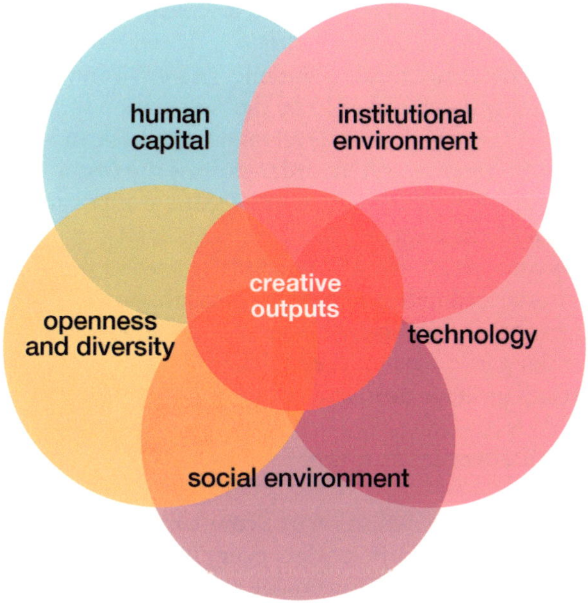

Figure 2.
The ECI six pillars (source: [39]).

Despite the European ESSnet framework, research at the national level on the types of indicators is still not exhaustive to frame the phenomenon, as shown by the culture and leisure survey whose indicators mainly concern: Cinema places; Number of museums, theaters, and cinemas; and Demand of museums, theaters, and cinemas (measured with the number of visitors to exhibitions and performances). On the other hand, already from the definition of the issue, it is clear that the analysis dimensions should be expanded: the cultural offer favors social contacts, develops new interpersonal networks, and offers leisure activities. Culture and leisure also include associations and sports centers [40].

In Unioncamere and Symbola's research, a step forward in terms of measurement was made by dividing the cultural system into five macro domains, but analyzing the phenomenon mostly from an economic point of view: creative industries (architecture, communication, and design), cultural industries (cinema, publishing, videogames, software, music, and press), companies working with historical-artistic heritage (museums, libraries, archives, archeological sites, and historical monuments), those working with performing or visual arts and creative-driven companies, not directly related to the sector, but employing cultural professionals or using cultural heritage as input to increase the symbolic value of their products and their competitiveness [20].

In 2018, there was an increase in this type of companies, the sector counts 416,080, the number that accounts for 6.8% of the total Italian economic activities and produces a turnover of about 96 billion euros (4 billion more than in 2017). There are 289,792 companies directly linked to cultural and creative activities, of which 129,533 are creative industries, and 147,153 are cultural; the other creative-driven companies are closely linked to the "Made in Italy" sectors. Overall, the performance of the individual sub-sectors is the following: companies operating in the communication sector grew by 1.3%, those operating in the performing arts by 2.7%, those operating in the historical-artistic heritage by 4.9%, those in design by 2.1%, those in video games and software by 2.7%, and the only sector to show a negative result is publishing, which is down 2.0% compared to 2017 [20].

On the basis of this theoretical framework, it is clear that the Italian context needs to be declined within the complex dimensions and interrelationships that culture can trigger, generating social, economic, and environmental impacts and effects. A first attempt was made in 2010 by the Association for the Economy of Culture and Federculture, which have developed a synthetic index of creativity starting from the ECI reference [39], in order to draw the first atlas of Creative Italy. It is a first descriptive tool on the strong commitment to supporting youth creativity by the various Italian Municipalities [41].

From the above considerations, it highlights the need of defining a framework of dimensions, criteria, and indicators able to measure, monitor, and evaluate successful practices from a sustainability point of view (social, economic, and environmental), taking into account the multidimensional approach that cultural activities and cultural heritage need and express.

1.2 The key role of intellectual property for cultural creative enterprises

According to the Baumann definition of a company, we are living "totally liquid" with continuous transformations that affect every process and tool, changing our perception of reality [42]. Digital natives have a different perception of the world around them and inevitably perceive it according to the mechanisms of digital transformation. The concept of "material" has also changed: it is perceived differently by people who obtain information, data, images, and videos immediately with different multimedia tools. The contents today are available instantly, in different formats and usable by various devices, in the metro, at the supermarket or in a bank. We are active spectators of what Rifkin [43] called "Age of Access," which entails a profound transformation of our society and experiencing the relationships between us and with products. Moving with a playful dimension on the web does not lead the legislator to consider any activity lawful and to ignore the legal implications of the online circulation of materials. The user asks whether it is always lawful to share and use files, images, programs, contents, and texts found on the Internet with author copyright and what are the strategies offered by technology in protecting the intellectual property and economic exploitation of a creative idea.

Copyright refers to the legal institution that protects the product/service of intellectual activity by recognizing a series of rights (moral and property nature) to the original author of the idea. The exercise of these rights by the author allows him to remunerate himself for a limited period through the commercial exploitation of the work. Copyright was born with the creation of the idea.

The author of the business idea automatically has the exclusive right to use it, and he can authorize or refuse its reproduction, distribution, execution, or representation. The lack of authorization and explicit consent by the author does not allow the potential user to hold and disseminate the work, even if it is present on the network. When the copyright is violated, there are some penalties, both criminal and civil, depending on the type of violation committed.

Copyright, with the emergence of the combination between digital technologies and internet, has lost the territorial dimension; in a globalized world in which physical distances between states and geographical and economic obstacles are reduced, a circulation of business ideas is very fast. Texts, images, videos, and audio tracks are made public freely and, above all, without any kind of intermediation.

This rapid data circulation makes copyright protection more complicated and has weakened enforcement actions to detect infringements. Copyright was born in England in the sixteenth century [44], thanks to the spread of the first automatic printing machines. The copyright was intended as the publisher's right and not as a recognized right of the author. The crisis of the traditional copyright model, no

DOI: http://dx.doi.org/10.5772/intechopen.92115

longer able to protect the authors' needs thanks to the information shared online, has implemented the creation of new web solutions allowing authors more direct protection of their products or services.

This type of model is the "guarantee threshold" in which the author communicates in advance the economic amount necessary to carry out the work. This economic amount will be the threshold.

Only when the total of the guarantees collected reaches the established threshold, the intermediary will enter into a contract with the author of the work in which the accumulated guarantees will participate. With the total amount of money to carry out the work, each subscriber of the guarantee has to pay for the guarantee and after the intermediary will advance a part to the author, as written within the contract.

When the work is completed and made available not only to the subscribers of the threshold but also to people who have not signed the guarantee, the intermediary will pay the remaining amount due to the author.

In Italy, the legal system recognizes copyright and the creation of an intellectual work thanks to creativity; the rights belong exclusively to the author. The web has made it necessary to adapt copyright protection to meet new requirements. The network becomes the place to search these creative works when the user decides to use products/services, and he becomes a communication tool and not a "mass point." New transnational regulatory sources are added to the Italian law by defining a regulatory system articulated on three levels: International treaties, specifically the Berne Convention for the Protection of Literary and Artistic Works, and the two WIPO Copyright Treaty (WCT) and the WIPO Performances and Phonograms Treaty (WPPT) of 1996.

Thanks to the signing of the Berne Convention, the ratifying countries have committed the creation of a unique discipline for the copyright and, by virtue of the principle of assimilation of Art. 5, the original works from ratifying country must enjoy in the other countries the same treatment guaranteed by national law to their citizens, in addition to the minimum ones provided by the Convention. On January 1, 1995, the Agreement on Trade-Related Aspects of Intellectual Property Rights was adopted, attached now to the deed of establishment of the World Trade Organizations (WTO), ratified in Italy with the law of December 29, 1974, no. 747, and dedicated copyright and related rights.

The EU Directives and Regulations deserve particular mention, in particular the Directive 2001/29/EC of the European Parliament and of the Council of May 22, 2001 on the harmonization of copyright and related rights in the information society (implemented with Legislative Decree no. 68/2003), have multiplied and accelerated the technological development of communication vectors, from production to exploitation.

Furthermore, the recently approved directive by the European Parliament on digital copyright, which contains safeguards for freedom of expression, allows news creators and editors to negotiate the payment of fees with the giants of the web for the use of copyrighted content.

The open model tries to open everything that the copyright would have kept closed through the authorization, a priori, of the use of the work. The diffusion of the open model in the field of artistic works is ascribable to the Creative Commons model, which has allowed, in an intelligent way, the diffusion of licenses designed for all types of intellectual works. It was born thanks to some researchers from Cambridge and Massachusetts, supported by many intellectuals, located in 70 countries around the world, and Creative Commons is a noncorporate body profit. In Italy, Creative Commons is based in Turin, at the Naxa Center for Internet and Society at the Politecnico di Torino. The main objective of this project is to promote a worldwide debate on the management of copyright and the dissemination of legal and technological tools that affirm a new model in the distribution of cultural products. Creative Commons [45] uses existing copyright to free creative works and disseminate them in a "certain rights reserved" regime, as opposed to the classic "all rights reserved" type. A challenge represented by an image in which the Creative Commons model appears as an intermediate model between the classic one ("all rights reserved"), typical of traditional copyright, and the "no reserved rights" model, typical of the integral public domain or a sort of "no-copyright." Creative Commons licenses ("CC licenses") consist in the granting by the licensor, free of charge, to the licensee for a period equal to the applicable copyright, of an authorization to perform, respecting conditions that vary depending on the license used, the reproduction, distribution, communication, and making available to the public. The main peculiarity of CC licenses consists in the fact that the licensor grants the licensee free of charge, for the entire duration of the applicable copyright (or for 30 years pursuant to Art. 1573 of the Civil Code, in the event that it is preferred to trace the Creative Commons Public Licenses CCPL in the case of leases) the authorization to perform, subject to conditions varying according to the specific CCPL used, some of the acts that the copyright rules reserve for the rights holder, including the reproduction, distribution, and communication to the public. Creative Commons licenses offer six different sections of copyright ("Attribution"). Cultural heritage, as defined by law, is the subject of public law and interest; to be usable by the community, they meet copyright and property rights, in which there is the right of the respective owners to the availability and even economic use of these goods, mostly intellectual works, by the collectives for cultural purposes or third parties for profit or profit. The relationship between cultural heritage and the copyright is set by the code, which explicitly saves the discipline of copyright, thus sanctioning a sort of double track between the private and individualistic protection of intellectual works and the protection of cultural assets, responding to a collective interest.

Subsequently, the regime evolved and the notions, as mentioned, partly overlapped. Indeed, there is a legislative evolution that starts from the 1939 law on cultural heritage, which excluded "works by authors living or whose execution does not go back more than fifty years." We can, therefore, say that without the explicit consent of the author, it is possible to prohibit their use by third parties. We speak, in fact, of copyright as authorship and possession of essential rights that allow not only to dispose of the product as you see fit but also the possibility of exploiting it for your economic gain. An exception is represented by "fair use," which becomes a possibility for third parties to use a work without requiring prior and explicit consent from the owner.

The lawful use of creative work and its enhancement through gamification, in order not to incur infringements of copyright, in addition to the prior consent of the author, must exclude that the use is only for commercial and profit. However, there are some critical issues relating to the absence of a common regulation for all the States that avoid the reproduction of different applications and situations based on the legislative system considered.

2. Materials and methods: collaborative decision-making processes and gamification for cultural heritage enhancement

2.1 Complex social shared values in co-evaluation approach

The Faro Convention [46] explains two particularly significant concepts on the value of cultural heritage for society: cultural heritage and economic activity, related to the need to involve every individual in the ongoing process of defining and managing cultural heritage, and the need to raise awareness of the economic potential of cultural heritage itself. Hybrid entities [47] called to produce economic value for cultural heritage, need to build heritage communities around it through collaborative social innovation processes that no longer respond to the ordinary structure of business projects.

Those organizations, which make relationality one of the drivers of their "doing business," expand the social perimeter to the point of becoming a strategic asset capable of overcoming the dichotomy between economic and social dimensions, profit and nonprofit. They feed on sociality, relationships, social capital present in the territories and through their activity they regenerate and nourish it, so that it is the lifeblood of their entrepreneurship and that of the territory.

These organizational models are increasingly linked to the community and the territory as an essential engine of their actions characterized by systemic, collaborative, human-centered approaches. Indeed, the community and the territories become at the same time, the beneficiaries of the offer of social enterprises, and their co-producers [48]. Between profit and nonprofit, the new hybrid organizational models dialog with public and private, merging economic and social value into shared value [29], traditional enterprise and social enterprise, producing communities and connections [49] in real urban laboratories, where the value of gift and market, as the result of entrepreneurship and participation, are hybridized. For these organizations, the passage from a design vision to a procedural one, from the restitution of service to its co-production, from the spatial concession to the co-management of the heritage, from a generalist market to territorial strategies, requires an essential reflection on the ways of eliciting values [50], which are particularly significant when the "waste heritages" are at the center of the valorization process. With the terms "waste heritages," we identify those abandoned or under-utilized urban places or buildings, the forgotten intangible cultural heritages, which in some way deny citizens the right to participate in the process of defining and managing heritage [46]. For these assets, whose value in the absence of a market cannot be assessed with traditional economic measures, evaluation techniques, and the articulation of the decision-making process plays a central role. The value of waste heritages can be interpreted as a set of complex values [5, 51] deriving from social interaction through the active involvement of informal and/or formal groups that cooperate with to restore or enhance value in use and value independent of use [52]. The evaluation that tries to keep together the individual and the social/community dimensions [53] has, therefore, the need to explore in-depth the Complex Social Value concept (CSV) [12], considering the different points of view, the complexity of the process in terms of contexts, resources, approaches, scales, stakeholders, and the coexistence of tangible and intangible values, intrinsic and economic at the same time, connected to heritage and community, to heritage and its potential use.

In order to identify and assess the different values that characterize waste heritage, the Collaborative Decision-Making (CDM) processes [54, 55] can be considered as the suitable context where it is possible to involve community members and share knowledge about the decisions to be made and together agree on the approach and decision-making principles to activate. Although the difficulties encountered by this type of process (cultural and linguistic differences, different preconceptions, incompatible political, and economic orientations) [55], there are considerable advantages in their implementation in complex processes. Collaborative decision-making processes succeed in bringing out and resolving conflicts, certainties, ambiguities characteristic of community-driven processes [54], combining the valuable contribution of all parties involved in a decision that, through collaborative and creative thinking, in a team of heterogeneous expert and nonexpert evaluators, can provide innovative and shared choices strongly rooted in the needs of the local community. In the evaluation processes, the preferred solutions are those that fall within the field of public-private-social negotiation, allowing the interpreters of the processes to achieve the maximum benefits, thanks to win-win-win strategies, which reflect models typical of the circular economy [56–58]. The more the choice will be supported by a collaborative process, the more creative the alternatives will be and the higher the possibility that they will become operational, as they can manage conflicts and involve both "stakeholders (stakeholders) and asset-holders (resource bearers)" [59].

The real challenge, therefore, becomes that of succeeding in eliciting and evaluating, through deliberative techniques [60], complex social shared values related to cultural heritage. The objective for social enterprises is to demonstrate to produce economic value starting from intangible values, so that the latter can be a driving force in the valorization strategies and become decisive in the organizations' and territories' strategic choices.

In order to expand the internal dimension of new hybrid enterprises (change management, soft skills, labor productivity) and their external dimension (engagement/co-production, reputation, financing, negotiating power) [61], the measurement of impact through open innovation [62] processes is strategic. The theory of change expressed by the impact value chain [63, 64] represents a useful tool to produce and assess small changes in the short term that are reflected through existing systems and lead to significant long-term change and impacts [65] (**Figure 3**).

The impact value chain [66–68], following the definition of the inputs, that is, the tangible and intangible resources available to the organization, consists of two main phases. In the first one "performance measurement," the activities and outputs (outcomes in terms of goods or services generated by the activities) are defined; in the second one "impact measurement," the outcomes (medium-long term effects generated by the outputs) and the impact, defined as a long-term sustainable change in people's conditions or the environment, are determined. At the European and global levels, there are currently several methodologies and tools for the measurement of outputs, outcomes, and social impact, which reconfigure the general structure according to its characteristics, territories, and communities of reference, identifying appropriate criteria and indicators [69].

Through this and other integrative tools, evaluation is not intended as a uniform metric but as a shared process built with and for the community [61, 70], like a participatory act that overturns the ordinary one-way evaluation paradigm,

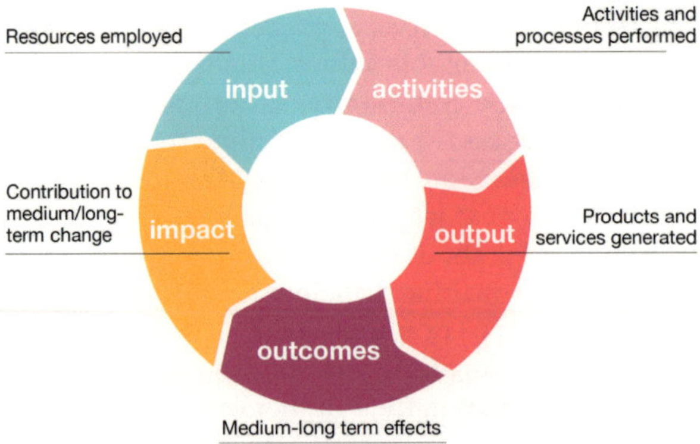

Figure 3.
Proposal for a new impact value chain (authors' elaboration) (source: [66–68]).

integrating evaluators and stakeholders and different skills in the process. The co-evaluation, more widespread in the teaching fields [71], in these complex processes is thus outlined as a systemic application of the evaluation process [72] that concerns a common good, experimenting the ability to evaluate together, each with its own different skills, to reduce conflicts and bring out common benefits, a tool capable of including the stakeholder, better defined as an "asset-holder," in the whole project process and at the same time evaluation.

By borrowing the application in the field of didactics, co-evaluation allows the user to learn from the activities of others, and developing the ability to discern between positive and negative aspects, to identify ways to improve, increase, and explore individual and collective opportunities, contributing to the user's empowerment included in the process. The action, connected to the evaluation through the whys and how, goes beyond the individual expression of a judgment, rather encouraging cooperation and the genesis of relational values and shared preferences, associated with interpersonal relationships and relation articulated by political and social norms [73].

According to the above perspective, the approach of co-evaluation is particularly significant in order to configure the structures of hybrid enterprises internally and to measure the social and cultural impact generated in actions for enhancing all intangible and tangible forms of cultural heritage.

2.2 A game-thinking-centered approach for creative cultural enterprises

The frontier of possible languages to build multi-group decision-making processes has widened very often through game thinking [74], and it is particularly relevant to analyze its potential for cultural organizations that activate processes of heritage enhancement through the formation of heritage communities. Indeed, the behavioral dynamics translated from the world of gaming bring different benefits:

- they facilitate the engagement of users with the activities of the company;
- they broaden how values are disseminated to the related cultural heritage, as well as the target audience;
- they allow to increase productivity, relationships, and co-production of goods and services through multi-stakeholder co-evaluation processes, facilitating the recognition of objectives and values, weights and preferences in a transversal way; and
- they offer enormous potential for the promotion of active and responsible forms of citizenship, as well as the co-design process promotion [75].

It is relevant to highlight the substantial differences between some of the different possible techniques:

- gamification: it is configured as the use of video games mechanics and dynamics within nongaming contexts, to create engagement, loyalty, improving a process or solving a problem [76]. More simply, "it is the use of game design elements in non-gaming contexts" [77]. Thus, a series of schemes experimented in videogames are used to invite the user to perform specific actions, make choices, to reward him in an intrinsic (triggering emotions in it) or extrinsic (points, prizes, etc.) way;

- pontification: it is a subcategory of gamification [78] and uses a basic points/reward scheme to move the user from point A (personal sphere) to point B (company sphere of interest);

- serious games [79]: they are virtual environments designed to develop skills and competencies transferable to the real world [78]; and

- serious urban games: they are playful and organized offline practices that take place in urban environments with some kind of technological/digital support, which serve social and cultural purposes [75] and transform urban spaces into a sort of playful interface.

Recent successful experimentations, one of which is the "Father and Son" project by Fabio Viola, for the National Archeological Museum of Naples [80] have shown how the application of game design to the experiences of heritage enhancement can be particularly appropriate not only in terms of engagement and dissemination, but also in terms of the reputation of the organization. After 11 months since the launch of the game "Father and Son" on digital stores, 2 million downloads have been exceeded by an international audience with an average age of 30 years, of which less than 10% are Italian. More than 18,000 people from all over the world visited Naples and the Archeological Museum unlocking the additional content of "Father and Son," and in general, the perception of Naples and the Museum, analyzing data from tens of thousands of reviews, has increased significantly thanks to the game. It should come as no surprise that, according to Google data for 2017, Naples is the most sought-after city on the search engine, while the MANN recorded for the first time in its history the goal of 500,000 visitors in 2017, with a growing trend [81].

From the virtual to the real dimension, from storytelling to storydoing [82], the world of the game offers new perspectives and ways of interaction to know, modify, evaluate, and influence people's behavior through their active involvement in processes, in the cultural sector too. A game is a system of interesting choices [83] and highlights one of the fundamental aspects drawn from the logic of gaming: that of structured choices in front of which players are led to reflect on what they have done and whether the best choices have been made. Brice Morrison designs the architecture of structured choices starting from four main pillars [84], for which the choice must be conscious, that is, critical thinking; the bearer of consequences, with immediate or long-term impact; comparable, the choices have a more significant impact than those of other people according to a psychological principle called "social influence." Finally, structured choices are permanent choices, that is, those that induce the user to consider a weight that cannot be erased, stimulating the search for the best possible solution [85].

In this way, the dynamics of the evaluation process find new conditions around which to structure itself and different times to implement, if we compare the ordinary timing of an evaluation process with those suggested by the game world, which has always accustomed the user to a constant and instantaneous evaluation [86]. Therefore, take into account the substantial difference between a game, designed to involve the user with his emotions, and real everyday life, which is often not structured with people's emotional involvement in mind. To this end, the use of intrinsic logics and behavioral structures, mechanics (product framework generally associated with some actions that allow achieving business objectives), and dynamics (needs and desires, studied by behavioral psychology, which are rooted in people) deduced from the world of the game becomes particularly effective for organizations engaged, through the enhancement of tangible and intangible cultural heritage, to build communities through collaborative processes.

Starting from the outlined process, the research aims to build a methodological approach (**Figure 4**) for rethinking the social enterprises as a creative cultural enterprise engaged in the reuse and enhancement of cultural heritage, exploring the role that game thinking can assume for them.

By identifying the thought of the game as a toolbox with technical and psychological, mechanical, and dynamic tools extracted from the game world, we investigate how it can be considered functional to the development of hybrid enterprises that aim and overcome globalization through strong territoriality.

A field of action of these organizations is identified in the intersection between the territories and the communities, and more specifically between the cultural heritage, tangible and intangible, and the heritage communities that surround it. The circularity between these two dimensions defines a symbiotic relationship between people, places, and traditions thanks to the relationships. Social enterprises engaged in the cultural field cannot disregard the relational and social character; on the contrary, it is precisely the social impact that represents their positioning element. If a generic capital company interfaces strictly with the value and economic impact of its actions, the social enterprise, not only aimed at the product as much as at the processes, has to confront itself with the production and evaluation of complex values because its impact is not only economical but also, and above all, environmental, social, and cultural.

In this perspective, it is of crucial importance to rethink evaluation techniques that can take into account these complexities, social innovation in the participatory component, and multi-stakeholder approaches that characterize the production of the collaborative enterprise.

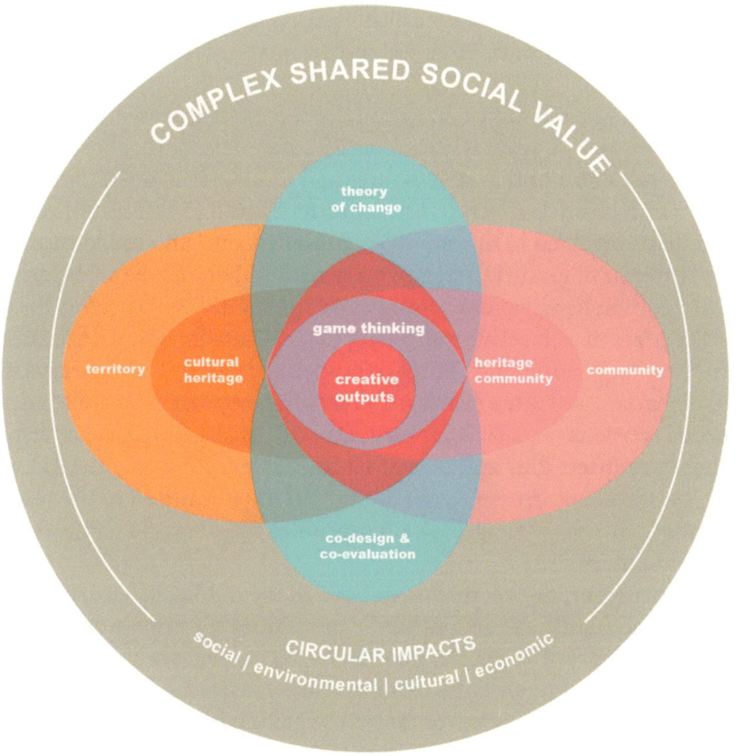

Figure 4.
A game-thinking-centered approach for creative cultural enterprises in the field of cultural heritage enhancement.

By co-evaluation, we mean a multi-group evaluation approach across the entire production process, in which everyone, with their skills, overcomes individual judgment to achieve common benefits by planning change together. To this end, the methodology of change theory expressed through the impact value chain is particularly significant for social enterprises. It is an approach to design and planning through a participatory process that allows the planning and evaluation of the change produced. The methodological approach thus defined places "game thinking" (**Figure 4**), with all its declensions, central in the internal and external development of the company, as it offers languages and behavioral structures in engagement, co-production, and social influence practices useful to polarize the field of action, through gamification, serious, and urban game. In social, cohesive, relational, and territorial enterprises, where processes and collaboration are the real challenges to change in value, the rules and languages transformed by the game world, by their human-centered nature, are significant to facilitate and simplify co-design and decision-making processes.

3. Results

3.1 The Play ReCH platform

The "San Sebastiano Monte dei Morti" Living Lab (SSMOLL) activated in 2018 from Blam in Salerno, in the South of Italy, through the adaptive reuse of the former sixteenth-century church abandoned for over 30 years, releases significant data about the perception of culture in the city of citizens.

Questionnaires and interviews carried out in the co-exploratory phase of the SSMOLL project claim that 43% of respondents expressed a lack of an integrated cultural offer and 75% dissatisfied with the quality of the offer and the poor communication; on the other hand, 90% see culture as the starting point for a process of urban regeneration.

Among the direct beneficiaries, based on the real experiences of SSMOLL, are schools and universities (5/year) with at least 25 students/month, artists and artisans (70/year), citizens (150/month), associations and foundations as partners already involved (20), with positive effects in economic terms and visibility. Indirect beneficiaries are the tourists involved in actions of "community tourism" and local authorities, facilitated by projects of cultural heritage valorization that include an expansion of the offer and a comprehensive communication campaign. The indicators for monitoring the project concern: subjective well-being (e.g., % satisfaction with leisure time), creative economy (e.g., number of people employed in cultural and creative work), cultural participation, and attractiveness (% satisfied population of cultural services in the city); social relations (e.g., % voluntary activity), social and civic participation.

From these premises, Play at Reuse of Cultural Heritage (Play ReCH) was born, a project of the social enterprise Blam, organization committed to the reuse and enhancement of cultural heritage, which promotes the idea of a game-thinking-centered hybrid enterprise.

The virtual platform, not yet made available to the public, offers, through a logic of gamification and pontification, cultural experiences of heritage enhancement structured on the model of serious urban games to connect, through social innovation processes, the virtual world with the real one thus transformed into a sort of playful interface. The main objective is to transform the citizen into the protagonist of his or her territory, leading him or her to face interactive cultural experiences transformed into "missions" able to break down the boredom found in the common standardized offer and to make the under-used heritage places more attractive and interactive.

Starting from the methodological approach, previously described, Play ReCH is configured both as a creative technological and as a real-life output.

The digital platform becomes the tool through which activating real actions, defined "missions", to produce circular impacts on the territories eliciting and evaluating the complex social values shared by communities through collaborative decision-making processes. Within the Play ReCH value chain model, the Blam organization divides the actions into five macro phases (**Figure 5**): (1) create a heritage community, (2) explore the cultural heritage, (3) co-design the Play ReCH missions, (4) co-evaluate the Play ReCH missions, and (5) generate the circular impacts. Each of these steps corresponds to the actions required by the users, intended as co-producers of the value chain. Taking into account the described actions, the interactive technology platform is designed. The citizens become protagonists of their city and "human sensors" of urban problems related to the cultural heritage enhancemente through the selection of experiences/missions.

1. "Create a heritage community". To understand how people interact with their cities and cultural heritage, and therefore to customize the cultural offer on target users, the social enterprise needs to know it in depth. The social enterprise added value is ensuring that individual user-stakeholders enter into relationships with others, thus generating a relational community and a stakeholder-map. To achieve this goal, Play ReCH adopts some game rules in the moment of user registration, where in addition to building their profile with personal data, the citizen releases other sensitive data relating to their interests and their territory. By building his or her character through the use of avatars and other engage devices, the player turns into a real urban sensor within a storytelling process in which he or she becomes the protagonist and collaborates with a community of friends and strangers to reach the final goal: to solve the problems of contemporary cities and a disused or abandoned heritage.

2. "Co-explore the cultural heritage". Starting from the methodological approach typical of living-lab to build collaborative processes based on the phases of

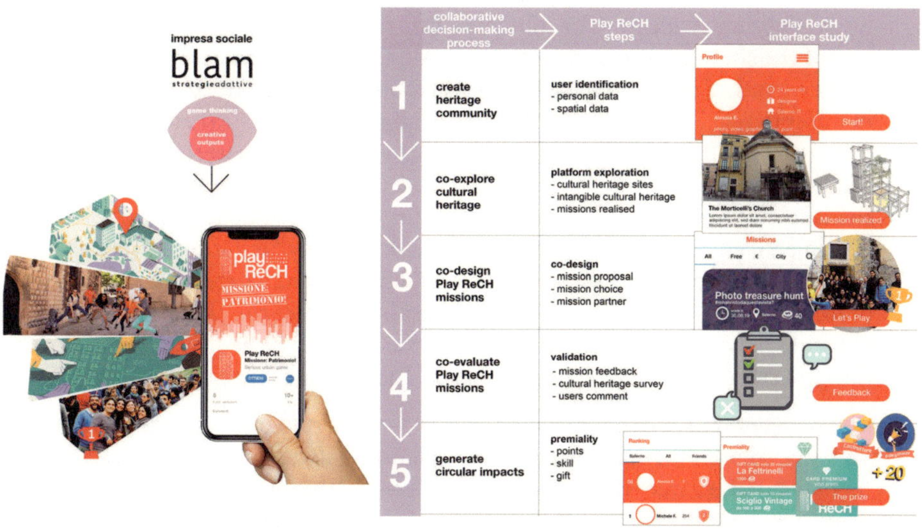

Figure 5.
The play ReCH (reuse cultural heritage) process.

co-explore, co-design, co-test/co-evaluate, Blam deeply analyzes the territories with communities, through collaborative processes. On the one hand, the Play ReCH platform leads the user during the registration phase to express his or her connection with the territory for establishing different starting perspectives; on the other hand, it stimulates the player to create deep links with the territory and with disused or under-utilized heritage. The co-exploration phase is therefore crucial, and it is expressed in the platform through the possibility of knowing the places interested by the single mission, or suggesting cultural heritage places in which the missions will be active. By choosing each site-specific mission, it is thus possible to view a digital showcase with a detailed description of the asset concerned, whose textual and photographic content can be produced collaboratively with experts in the sector, local realities, local associations, thematic social groups, owners or managers of the assets, and so on. The co-explore and community user profiling phase allow defining an input component useful to the value chain of the company's impact.

3. "Co-design Play ReCH missions". The experiences' design phase to enhance the cultural heritage, tangible and intangible, is the moment when the main goal of Play ReCH is synthesized into sub-goals, focused on specific places (in the case of site-specific missions, e.g., in a city, neighborhood, museum, etc.), or on specific experiences that can be replicated in different "clusters" of places (in the case of site-specific missions where the required actions can be carried out in all territories, e.g., in all museums, in a favorite cultural heritage place, etc.). The first type of mission becomes relevant as a connection between the enterprise and the communities in the territories where regenerative strategies are being carried out. The second type of activity, on the other hand, represents the moment when the enterprise becomes scalable, thanks to the technological infrastructure that finds a global level in which to bring out the strong territoriality of its strategies, as well as the realization of a network between people and places. In this framework, the co-design activities of the impact value chain have a key role within the collaborative decision-making process activated. The registered user can choose among the available "missions" in which activate the place or the skills required.

   Each experience releases the following information: (1) definition of the problem of the single mission; (2) determination of the objective of the challenge to be faced and of the time limit for its realization; (3) regulation to be respected in order to reach the objective; (4) knowledge of the physical and theoretical context of what is faced in the mission (through digital material disseminated on the platform), in order to provide tools and information useful to increase the user's capabilities; (5) points and credits that can be acquired by reaching the objective; (6) possibility to invite friends to "play" together for collaborating as a team; and (7) skills to collect by facing the challenge. The skills, interpreted in the game as character medals or abilities, are the skills required by the player to participate in the mission, or those that can be acquired at the end of the experience, thus leveraging the users' "achievement" feature.

   The registered user can also view the missions carried out so far geo-located on a shared mission map. For each experience, he/she can know the places involved in the game, the partners who have collaborated in the initiative, the users who have carried it out and the results obtained; the player can propose a mission by suggesting it on the platform through the completion of the seven fields previously identified. Moreover, it is possible to indicate places in the city where the player wants to intervene. In this way, the digital platform also becomes a repository of the various voluntary data collected, encouraging the direct involvement of citizens and making them protagonists of the intervention choices.

The experiences' co-design can take place with the different actors of the city: from citizens, administrations, local organizations to the cultural actors of the territory (artists, associations, museums, foundations, galleries, etc.). In this way, the platform concretizes and shows the networking work that a social enterprise engaged in the construction of territorial strategies. The app and the website thus become open contexts for stakeholders who collaborate in the construction of the missions who promote their implementation and who come into contact with other realities on a national scale.

Storytelling and the variety of media needed to solve the challenges become fundamental to involve the citizen in a continuous role play.

4. "Co-evaluate Play ReCH missions". The mission performance in real life is crucial because of territories, and communities meet and create links in real life. The validation of the experiences carried out, according to the receipt of the award, becomes a critical issue both for the game and for the choice of evaluation methods. In order to co-evaluate personal experiences and their impact on the territory, Play ReCH requires the user to provide ex-ante and ex-post feedback on the acceptance of the challenge, to analyze the variation in knowledge of the places involved in the mission and the degree of satisfaction of the completed experience compared to the initial expectations placed in it. Besides, Blam favors users' ability to work together for defining creative solutions to urban problems. Therefore, the platform favors the possibility to comment, in a structured and constructive way, the actions carried out by the other players and to evaluate the comments received, in order to start virtuous debates and produce links between the participants. In this way, the registered player will be able to view the open profiles of other users and their content, and the missions carried out, the media produced, the points and skills acquired.

   Co-evaluation assumes a strategic role in the impact value chain under construction, since it allows, through a basic set of criteria and indicators, to renew and assign weights in order to measure the outputs of the single experiences concluded, that is, the products and services generated by the activities, as well as, in the medium-long term, the outcomes produced by "clusters" of actions.

5. "Generate circular impacts". In order to contribute to building social, cultural, and economic networks and to generate circular impacts, Play ReCH activates gamification processes with related scores, skills, and awards. Through a premium fidelity card, the registered user can also accumulate credits necessary to win real prizes. Play ReCH cultural partners, in the "prizes" section of the digital platform, dedicate gifts and bonuses to premium players divided by credit bracket. The more expensive prizes, the higher the number of credits. The player can choose which rewards collect based on its score.

Play ReCH prefers the distribution of prizes useful to increase the player's capabilities: cultural experiences, museum access, workshop activities, training courses, eco-design products, and so on.

User-experience and engagement dynamics become fundamental in this game-thinking-based ecosystem structured as follows. Specifically, Play ReCH is built around the mechanics and dynamics of the game, using the techniques of gamification for the construction of the platform and its sub-category "pontification" for the awarding of points, skills, prizes, etc.; the model of serious games for the realization of the mission purely online; and the structure of serious urban games for the missions' design spread thanks to the platform.

Blam's final goal is using Play ReCH data and considers reality as a playground in which building heritage communities that share structured regeneration strategies, for identifying new ways of evaluating the complex shared social values and the effects generated in the impact value chain.

The experiences are designed according to the serious game model used by Jane McGonigal in his "Evoke" experiment [87]:

- definition of the problem of the single mission;
- determination of the objective of the challenge to be faced;
- regulation to be followed to achieve the goal; and
- knowledge of the physical and theoretical context of what is faced in the mission, in order to provide tools and information useful to increase the user's capabilities.

The mission, which takes the player to the real places of the territories, puts the user in contact both with a community of people and with places of cultural heritage probably never known before.

Experiences activated in Salerno demonstrated, during the day dedicated to a site-specific mission that involved a city museum with an average number of daily users equal to 10, how the flow of visitors increased up to 60 in a single afternoon.

The mission validation, through feedback from the player, triggers the assignment of the score and then the repositioning in an overall ranking where you can compare with all other users or join the team with them. The score also increases according to the ability to share one's experiences and comment constructively on other users' contributions, thus encouraging the citizen not only to achieve results by producing creative answers to the problems of contemporary cities but also to collaborate in order to identify possible shared solutions. The construction of relationships and sociality is rewarding values compared to the quality of individual proposals [88].

Indeed, communities are involved through gamification as an online and offline co-production process for the creation of cultural and creative services coming from the bonds generated in the network.

The Play ReCH players can always be informed and become co-producers of a specific action to enhance heritage by taking part or proposing new experiences, in the logic of healthy competitiveness that stimulates collaboration and innovation. The data collection of the players and their actions helps the continuous re-definition of the value proposition of this social enterprise.

Through a premium subscription, which it is possible to access through the purchase of a fidelity card, the player accumulates credits and access prizes made available by cultural partners. The aim is to provide real rewards aimed at increasing players' capabilities, in terms of training, experience, and creativity.

Play ReCH is configured as the technological infrastructure of the project. Based on integration and interactivity criteria, it is a concrete support tool for the digital strategy. The web/mobile platform will be developed in-house with Open Source technologies to ensure full management autonomy in the long term. It will contain valuable Call to Action, Gamification, Geolocation, and Data Collection modules to guide the project strategies. An E-Commerce section will be dedicated to the purchase of cards and premium activities. It will integrate fb + ig social platforms for mass sharing, participation in gaming mechanisms, and social marketing operations to amplify engagement.

In this way, the enhancement of heritage becomes not only conservation and transmission of cultural heritage to future generations but also a continuous experience of re-discovery and re-writing by experts and non-experts, in a dynamic participatory process called to constantly elicit and co-evaluate emerging complex values. Play ReCH thus aligns itself with the 2005 Faro Convention, which establishes how cultural participation has a significant impact on people's quality of life.

3.2 Toward a new model of creative cultural enterprise for cultural heritage enhancement

One of the primary purposes of Play ReCH is to implement an integrated cultural offer and accelerate the spread of virtuous behavior in the reuse and enhancement of cultural heritage, generating new jobs in the creative field and forming an active, creative community. Belonging to a membership that gives access to dedicated services helps to strengthen the territorial identity and generate the empowerment. At the same time, the promotion of cultural and creative services supports local actors and improves the degree of user satisfaction with the existing cultural offer.

If in the short term it is possible to evaluate an immediate increase in the number of visitors, an increase in the monetization of the cultural heritage, and a more widespread recognition of the goods affected by the experiences, in the medium to long term, it is possible to see how the degree of involvement in the consumption of cultural experiences allows to strengthen the link with the territory, produce greater social cohesion, and contribute to the reduction of cultural poverty.

MarketsandMarkets is showing a significant growth trend for the use of applications, platforms, and methodologies based on the logic of the game, going from a turnover of 1.65 billion in 2015 to over 11 billion dollars in 2020. Virtual reality and game thinking, in general, are now closely connected to the business, able to consolidate business relationships with existing customers and create new ones with those not yet loyal. The #PlayOriginal campaign launched by Original Marines in 2015, which stimulated customers to face offline and online missions at the stores in order to earn bonuses and awards, in just 4 months has recorded 300,000 visits to the site, 30,000 subscribers of which 85% from the web and 15% from stores, 10,000 photos received to participate in the missions, 10 million impressions of the campaigns, and more than 2000 sales generated in stores thanks to the contest [89].

In the field of cultural heritage, the application of game design mechanics and dynamics is still not very widespread. It is still considered by organizations that see its potential only in experiences intended as an output of the organizations themselves, as concluded packages that emerge with a discontinuous offer (on commission) and with a little interactive technology (web site and social page). In 2015, the experience of CriticalCity Upload also ended, an Italian non-profit organization, which in 4 years has promoted experiences of urban creative transformation and civic activism based on fun theory, co-designed or totally proposed by users through an interactive web platform where it was possible to propose missions, take part, accumulate points and climb the ranking.

Since 2011, thanks to the support of foundations and cooperatives, 1092 hours of game days have been accumulated, 21,064 missions carried out in Italy and around the world, reaching 13,901 international players.

From the analysis of competitors, Play ReCH Blam supports the idea of a hybrid social enterprise that integrates profit and non-profit to provide a continuous cultural offer thanks to an interactive technology platform/app.

To this end, a freemium B2B and B2C business model is structured. The core of the BM is represented by the potential franchising of the platform, which is the possibility to resell it to entities (municipalities, regions, schools, etc.) in order to create geographically divided substructures capable of disseminating site-specific missions for their territories: on one hand, if this guarantees the promotion and an unprecedented involvement of their tangible and intangible cultural heritage, and on the other hand, it guarantees loyalty from players, local or not, who intercede with it in order to overcome missions and earn points, learning and promoting solutions to their critical issues. Other interlocutors of B2B Play ReCH, with whom the player can co-design site-specific missions, are as follows:

- the cultural organizations in the area (museums, churches, archeological parks, etc.);

- cultural associations and creative people who can find in the platform a national showcase for their activities, and in Blam a team of experts to co-design cultural experiences with a high degree of involvement; and

- the commercial actors of the territories, which are part of the cultural missions with locations, products, and corporate mission.

Added to these are the Play ReCH cultural partners, who are those guarantee user rewards on a national scale (chains of bookshops, publishing houses, newspapers, training centers, etc.), and the other urban game designers, for whom Play ReCH becomes a marketplace with a commission percentage.

Through online/offline gamification tools, it is ensured that the process of knowledge and enjoyment of cultural heritage is able to generate regeneration "experiences" based on sustainability, equity, and protection of creative expressions. Considering the high unemployment rate in Italy between 15 and 34 years old (17.8% on a national scale and 16.2% in the South, with a figure almost double the national average, according to ISTAT data as at December 12, 2019) and more specifically the unemployment rate among artists and creative people in general (26.8% for those under 30 and 36.3% between 31 and 40 years old, according to Inps 2017 data), we want to overcome the gap with the use of the web as a multiplier of projects, capable of generating new forms of collaboration and training. An expected number of new employees in the social enterprise (among partners, working partners, and collaborators) of 6 is expected, where 5 are women from Campania, all under 35 of which 3 NEETs. It is hypothesized that, starting from these opportunities for contract work or consultancy, the project will be able to build new jobs in the cultural and creative field for actors and users involved.

The user-centric enterprise formulated in this way expands the rigid perimeter of the canonical closed and polarizing for-profit organizations concerning the urban and social context, considerably reducing its index of centrality to open up to the relational dimension in which, more than the top of relationships, it becomes the producer of polycentric relationships. The user is no longer just a stakeholder of interest, but an asset-holder of resources [47], co-producer of content, and co-evaluator of impacts. The ReCH players can always be informed and become co-producers of a specific action to enhance heritage by taking part or proposing new experiences, in the logic of healthy competitiveness that stimulates collaboration and innovation. The data collection of the players and their actions helps the continuous re-definition of the value proposition of this social enterprise.

Personal user registration, continuous feedback, choices, and proposals allow Blam to have a dynamic and continuously updated pool of information to better understand the relationship between citizens and their heritage and reformulate the offer in real time. With Play ReCH, citizens become urban sensors that bring and produce value.

4. Discussion and conclusions

The Country Brand Index 2018 highlights that Italy is the first country in the world in terms of cultural influence. In the same year, the Federculture statistics include Italy among the worst in the world in terms of cultural heritage valorization. This paradox emerges in the 2018 European Year of Cultural Heritage, on the real opportunities for reuse and enhancement of Italian Heritage, if it is true that all museums, monuments, and national archeological sites bill in a year as much as one French museum, the Louvre [90].

All these translate into two main critical factors: the loss of earnings for owners and managers of a vast under-utilized asset on the one hand, and the continuing growth of educational poverty in Italy on the other hand.

The Play ReCH web-platform, winner of the 2019 Welfare Che Impresa call, can be the tool for rethinking the management model of cultural heritage reuse through gamification processes in combining technology and reality, involving city users within creative processes co-designed with different local actors.

The experience on the Play ReCH platform starts with profiling to become part of the community, releasing at the same time necessary information to customize the offer. Starting from an overall macro-mission that from the beginning puts the user in front of a global challenge, different issues related to the cultural, social, economic, and environmental crisis of the territories, expressed in daily missions, site-specific experiences, or generically addressed throughout the country, are declined. Each mission has a precise time to be accomplished, releasing skills based on the skills to be put into play or to be acquired with the specific experience, and points useful to climb positions in a general ranking.

The networking of community-actors-places accelerates the production of regenerative actions, thus transforming fragile urban heritage into catalysts of new energies. Indeed, by purchasing the Play ReCH card, the various players, members of the network, access the dedicated services, accumulate the points and receive the prizes, participating in creative cultural experiences useful for enhancing the cultural heritage and implementing the urban regeneration processes.

The project aims to respond to the social need of culture as welfare demand, identifying young people and cultural heritage, disused or under-used, as crucial resources for triggering new projects. Through the online/offline gamification tools, the process of knowledge and valorization of cultural heritage can generate “experiences” of urban regeneration, based on sustainability and protection of creative expressions, overcoming the gap with the use of the network as a multiplier of projects, able to generate new forms of collaboration in the game. The Play ReCH project has the potential to redesign the relationship among territories and entrepreneurship starting from processes of hybridization and contamination between profit and nonprofit, between formal economy and new economies, and leveraging the potential provided by technologies and gamification processes. The Play ReCH platform, for better understanding social complex dynamics to consider, intends experimenting the proposed approach in different territories in order to test how gamification can improve a real cultural valorization process.

The proposed methodological approach allows answering the first of our research questions by identifying how it is possible to build a creative cultural enterprise for cultural heritage enhancement able to implement innovative evaluation and management models in terms of the business value proposition.

At the same time, experimentation with the Play ReCH platform represents the way to test how gamification can improve collaborative decision-making processes for the enhancement of cultural heritage.

Through the analysis of the methodological approach, it is possible to highlight how some issues remain open which only direct experimentation will allow to explicitly explain, by reviewing the articulation of the phases and the ways of implementing the game.

If we analyze the first phase, "Create a heritage community," it is possible to highlight that it represents a real opportunity to activate and make protagonists, citizens who progressively establish relationships and ties with the cultural asset, becoming part of the same community, in coherence with the indications of Faro's declaration on heritage communities.

The second phase, "Co-explore the cultural heritage," allows promoting collaboration between the different players, facilitating interaction and allowing to identify shared missions capable of generating a chain of values and impacts that affect both the community and the assets.

The third phase, "Co-design Play ReCH missions," allows to identify the missions and select the preferable actions. This phase represents the most complex moment of the decision-making process, in which it becomes essential to be able to define the feasible missions, shared and able to improve the conditions of cultural heritage without compromising their characteristics. In this phase, the management of the game is crucial for the outcome of the missions.

The fourth phase, "Co-evaluate Play ReCH missions," clearly introduces a collaborative evaluation process, which allows the different players not only to self-assess their missions and their results but also to evaluate those of the other players. In this phase, players' awareness of the missions' objectives is fundamental, respecting and valuing the other players. Furthermore, how co-evaluation is carried out is decisive, which must be open, democratic, and inclusive.

The fifth phase, "Generate circular impacts," includes the ability of each mission to determine impacts on the cultural heritage and the territory consistent with the principles of the circular economy. In this phase, it is possible to verify the validity of the mission, its ability to influence the behavior of the players and to involve new players, to activate a chain of values and positive impacts. It is not evident that missions can generate circular impacts and the selection of actions that are capable of doing so can be one of the criteria to be taken into account in identifying missions.

By observing the stages of the process, it can be highlighted how the role of the cultural and creative enterprise that manages the Play ReCH platform is determined and requires skills both on the enhancement and management of cultural heritage and on multidimensional assessment, processes focused to conflict management.

The application of the Play ReCH platform will allow to test the decision-making model focused on game-thinking and to verify in operational terms how much the gamification process will contribute to improving strategies for enhancing cultural heritage.

Acknowledgements

Within the unitary work, Gaia Daldanise developed a literature review on Cultural creative production evaluation/monitoring frameworks (Section 1.1) and

Paolo Di Lauro developed a framework about intellectual property regulations for creative business ideas (Section 1.2). Ludovica La Rocca developed materials and methods on collaborative decision making, gamification, and co-evaluation (Section 2), and Maria Cerreta developed the introduction (Section 1) and the discussion and conclusions on the whole process (Section 4). All authors developed the results of the Play ReCH experiment (Section 3).

The Play ReCH is a Blam team project (https://www.blamteam.com/play-rech/) funded by the "Welfare Che Impresa!" grant thanks to the support of SNAM foundation.

Author details

Maria Cerreta[1*], Gaia Daldanise[2], Paolo Di Lauro[3] and Ludovica La Rocca[1]

1 Department of Architecture (DiARC), Federico II University of Naples, Naples, Italy

2 National Research Council of Italy (CNR), Institute of Research on Innovation and Services for Development (IRISS), Naples, Italy

3 Suor Orsola Benincasa University, Naples, Italy

*Address all correspondence to: maria.cerreta@unina.it

References

[1] United Nations. Transforming our World: The 2030 Agenda for Sustainable Development. United Nations [Internet]. 2015. Available from: https://www.refworld.org/docid/57b6e3e44.html [Accessed: 07 January 2020]

[2] Bosi G. L'impresa culturale: Diritto ed economia delle comunità creative. Bologna: Il Mulino; 2017

[3] Solima L. L'impresa culturale: processi e strumenti di gestione. Roma: Manuali Universitari, Carocci editore; 2005

[4] Bertacchini EE, Bravo G, Marrelli M, Santagata W. Cultural Commons: A New Perspective on the Production and Evolution of Cultures. Cheltenham: Edward Elgar Publishing; 2012

[5] Cerreta M. Thinking through complex values. In: Cerreta M, Concilio G, Monno V, editors. Making Strategies in Spatial Planning: Knowledge and Values. Dordrecht: Springer Science & Business Media; 2010. pp. 381-404

[6] Forester J. Beyond dialogue to transformative learning: How deliberative rituals encourage political judgment in community planning processes. Evaluating Theory-Practice and Urban-Rural Interplay in Planning. 1997;**37**:81-103

[7] Scott AJ. Creative cities: Conceptual issues and policy questions. Journal of Urban Affairs. 2006;**28**(1):1-17

[8] Evans G. Measure for measure: Evaluating the evidence of culture's contribution to regeneration. Urban Studies. 2005;**42**(5-6):959-983

[9] Miles S, Paddison R. Introduction: The rise and rise of culture-led urban regeneration. Urban Studies. 2005;**42**(5-6):833-839

[10] CHCfE Consortium, Cultural Heritage Counts for Europe. 2015

[11] Boschma R. Proximity and innovation: A critical assessment. Regional Studies. 2005;**39**(1):61-74

[12] Fusco Girard L, Nijkamp P. Le valutazioni per lo sviluppo sostenibile della città e del territorio. Vol. 74. Milano: FrancoAngeli; 1997

[13] Cerreta M, Daldanise G. Processi decisionali innovativi per la valorizzazione del patrimonio culturale: le imprese culturali e creative sostenibili. In: Genovese RA, editor. Patrimonio e città storiche come poli di integrazione sociale e culturale, sostenibilità e tecnologie innovative. Napoli: Giannini Editore; 2018. pp. 201-220

[14] Osterwalder A. The business model ontology: A proposition in a design science approach [thesis]. Switzerland: University of Lausanne; 2004

[15] Perna R, Cavicchi A, Cutrini E, Fidanza B. Playmarche: un distretto regionale dei beni culturali 2.0/Playmarche: a 2.0 regional cultural heritage district. Il Capitale Cultural: Studies on the Value of Cultural Heritage. 2015;**3**:99-136

[16] Council of Europe, Council of Europe Framework Convention on the Value of Cultural Heritage for Society [Internet]. 2005. Available from: https://www.coe.int/it/web/conventions/full-list/-/conventions/treaty/199 [Accessed: 10 January 2020]

[17] Clemente M, Giovene di Girasole E. Friends of Molo San Vincenzo: Heritage Community per il recupero del Molo borbonico nel porto di Napoli. In: Pavan Woolfe L, Pinton S, editors. Il valore del patrimonio culturale per la società e le comunità, la convenzione del Consiglio d'Europa tra teoria e prassi. Venezia: Linea Edizioni; 2019. pp. 173-189

[18] Calvaresi C. Community Hub, due o tre cose che so di loro. cheFare - cultura

e innovazione [Internet]. 2016. Available from: www.che-fare.com/community-hub-due-o-tre-cose-che-so-di-loro/ [Accessed: 10 January 2020]

[19] Throsby D. Economia e Cultura. Bologna: Il Mulino; 2005

[20] Unioncamere-Fondazione Symbola. Io sono Cultura—2019. L'Italia della qualità e della bellezza sfida la crisi [Internet]. 2019. Available from: https://www.symbola.net/ricerca/io-sono-cultura-2019/ [Accessed: 11 January 2020]

[21] Valentino PA. L'impresa culturale e creativa: Verso una definizione condivisa. Economia Della Cultura. 2013;**23**(3):273-288

[22] DCMS. Creative Industries Mapping Document. London, UK: DCMS London; 1998

[23] Santagata EW, Translation E, Kerr D. White Paper on Creativity Towards an Italian Model of Development. Citeseer. 2009

[24] British Council, European Creative Hubs Network. 2018 [Online]. Available from: https://www.creativehubs.eu/

[25] UNESCO, Convention on the Protection and Promotion of the Diversity of Cultural Expressions, Paris, [Internet]. 2005. Available from: http://portal.unesco.org/en/ev.php-URL_ID=31038&URL_DO=DO_TOPIC&URL_SECTION=201.html [Accessed: 10 February 2020]

[26] Fusco Girard L. The complex social value of the architectural heritage. Icomos Information. Naples: ESI; 1986. pp. 19-22

[27] Porter ME. The value chain and competitive advantage. In: Understanding of business process. New York: Routledge; 2001. pp. 50-66

[28] KEA European Affairs. The Economy of Culture in Europe. Brussels [Internet]. 2006. Available from: https://ec.europa.eu/assets/eac/culture/library/studies/cultural-economy_en.pdf [Accessed: 29 November 2019]

[29] Porter ME, Kramer MR. The big idea: Creating shared value. Harvard Business Review. 2011;**89**(1):2

[30] DelBaldo M, Demartini P. Responsabilità sociale di territorio, network sinergici e governo locale. Piccola Impresa/Small Business. 2016;**3**:54-81

[31] UNESCO. UNESCO Global Report on Culture and Sustainable Urban Development [Internet]. 2016. Available from: https://unesdoc.unesco.org/ark:/48223/pf0000245999 [Accessed: 10 March 2020]

[32] UNESCO. The 2009 UNESCO Framework for Cultural Statistics (FCS) [Internet]. 2009. Available from: http://uis.unesco.org/sites/default/files/documents/unesco-framework-for-cultural-statistics-2009-en_0.pdf [Accessed: 20 February 2020]

[33] UNESCO. UNESCO Culture for Development Indicators (CDIS) [Internet]. 2014. Available from: https://en.unesco.org/creativity/activities/cdis [Accessed: 20 January 2020]

[34] European Commission, European Statistical System Network on Culture, Final Report, European Commission—Eurostat [Internet]. 2012. Available from: https://ec.europa.eu/assets/eac/culture/library/reports/ess-net-report_en.pdf [Accessed: 10 February 2020]

[35] Cicerchia A. ESSnet Culture 2011: le definizioni dell'occupazione culturale. Economia Della Cultura. 2012;**22**(1):125-132

[36] European Commission. NACE Rev. 2. Statistical Classification of Economic Activities in the European Community. Eurostat Methodol. Working Paper. 2008

[37] European Commission. The Cultural and Creative Cities Monitor [Internet]. 2017. Available from: https://composite-indicators.jrc.ec.europa.eu/cultural-creative-cities-monitor/ [Accessed: 20 January 2020]

[38] Cerreta M, Daldanise G, Sposito S. Public spaces culture-led regeneration: monitoring complex values networks in action. The Journal Urbani Izziv—Urban Challenge. 2018;**29**:9-28

[39] KEA European Affairs. The impact of culture on creativity [Internet]. 2009. Available from: https://keanet.eu/publications/?term=The+impact+of+culture+on+creativity [Accessed: 21 November 2019]

[40] ISTAT. Cultura e tempo libero [Internet]. 2019. Available from: https://www.istat.it/it/files//2019/12/C10.pdf [Accessed: 10 March 2020]

[41] Cicerchia A. Economia della cultura e giovani: Dalle buone pratiche all'indice di creatività. Roma, Edizioni ComuniCare; 2010

[42] Bauman Z. Culture in a Liquid Modern World. London: Polity Press; 2011

[43] Rifkin J. The age of access: The revolution of the new economy. Barcelona Paid. Robson B, Lymperopoulou K, Rae A. 2008. People Move Explor. Funct. Roles Deprived Neighb. Environment and Planning A. 2000;**40**(11):2693-2714

[44] Deazley R. The statute of Anne and the great abridgement swindle. Houston Law Review. 2010;**47**:793

[45] European Parliament, Legislative Resolution of 26 March 2019 on the Proposed Directive of the Parliament and of the Council on Copyright in the Digital Single Market. 2019. Available from: https://eur-lex.europa.eu/legal-content/EN/TXT/?uri=EP%3AP8_TA%282019%290231

[46] European Council. Convenzione Faro. 2005

[47] Venturi P, Zandonai F. Imprese ibride: Modelli d'innovazione sociale per rigenerare valori. Milano: Egea; 2016

[48] Sturabotti D, Venturi P. Iris network. 2016 [Online]. Available from: http://www.rivistaimpresasociale.it/rivista/item/157-nuove-geografie-del-valore-e-imprese-coesive/157-nuove-geografie-del-valore-e-imprese-coesive.html?start=1

[49] Fusco Girard L. Nuovo umanesimo e rigenerazione urbana: Quali strumenti per una crescita economica inclusiva? In: Mazzinghi L, Spini A, editors. Per una nuova visione della città. Firenze: Polistampa; 2015

[50] Cerreta M, Izzo F. Napoli/Naples. Culture, Creative, Community Hub. Valori e sinergie della città storica in rete. Values and synergies network of the Historic City. Naples: Clean; 2016. ISBN: 978-88-8497-609-3

[51] Fusco Girard L. Creative evaluations for a human sustainable planning. In: Cerreta M, Concilio G, Monno V, editors. Thinking through Complex Values, Making Strategies in Spatial Planning. Dordrecht, The Netherlands: Springer; 2010. pp. 305-328

[52] Venturi P, Calderini M. Serve un nuovo paradigma di sviluppo per la rigenerazione urbana. Percorsi di secondo welfare. 2018 [Online]. Available from: www.secondowelfare.it

[53] Kajihara A, Harakawa T. Model of photovoltaic cell circuits under partial shading. In: Proceedings of the IEEE International Conference on Industrial Technology (ICIT '05); 14-17 December 2005; Hong Kong. New York: IEEE; 2006. p. 866-870

[54] Wilson MA. Collaborative decision making: Building consensus group decisions for project success. In:

Proceedings of the Project Management Institute Global Congress North Americas (2003). 2003. pp. 1-7

[55] Tisdall A. Collaborative decision making. In: Advanced ATM Symposium and Workshops. 2013

[56] Iraldo F, Bruschi I. Economia Circolare: principi guida e casi studio. Bocconi, Milan: IEFE; 2015

[57] Bompan E, Brambilla IN. Che cosa è l'economia circolare. Milano: Edizioni Ambiente; 2016

[58] Ellen MacArthur Foundation. Cities in the Circular Economy: An Initial Exploration. Ellen Macarthur Foundation [Internet]. 2017. Available from: https://www.ellenmacarthurfoundation.org/assets/downloads/publications/Cities-in-the-CE_An-Initial-Exploration.pdf [Accessed: 20 December 2019]

[59] Zandonai F, Venturi P. Dove: La dimensione di luogo che ricompone impresa e società. Milano: Egea; 2019

[60] Proctor W, Drechsler M. Deliberative multicriteria evaluation. Environment and Planning C: Government and Policy. 2006;**24**(2): 69-190

[61] Zamagni S, Venturi P, Rago S. Valutare l'impatto sociale. La questione della misurazione nelle imprese sociali. Impresa Sociale. 2015;**6**:77-97

[62] Chesbrough H, Vanhaverbeke W, West J. Open Innovation: Researching a New Paradigm. New York: Oxford University Press on Demand; 2006

[63] Weiss CH. Nothing as practical as good theory: Exploring theory-based evaluation for comprehensive community initiatives for children and families. New Approaches to Evaluating Community Initiatives: Concepts, Methods and Contex. 1995;**1**:65-92

[64] Taplin DH, Clark H, Collins E, Colby DC. Theory of Change Technical Papers: A Series of Papers to Support Development of Theories of Change Based on Practice in the Field. New York: ActKnowledge; 2013

[65] Dees JG. The Meaning of Social Entrepreneurship. Standford University: Draft report for the Kauffman Center for Entrepreneurial Leadership [Internet]. 1998. Available from: https://centers.fuqua.duke.edu/case/wp-content/uploads/sites/7/2015/03/Article_Dees_MeaningofSocialEntrepreneurship_2001.pdf [Accessed: 15 December 2019]

[66] Carlucci D, Marr B, Schiuma G. The knowledge value chain: How intellectual capital impacts on business performance. International Journal of Technology Management. 2004;**27**(6-7):575-590

[67] Vesce C. Analisi, misurazione e valutazione dell'impatto sociale delle Reti di Economia Solidale. Economiasolidale.net [Internet]. 2016. Available from: http://www.economiasolidale.net/sites/default/files/allegati/2016_SELS_07_ImpattoSocialeRES.pdf [Accessed: 14 December 2019]

[68] Kail A, Lumley T. Theory of change: The beginning of making a difference. London: New Philanthropy Capital. 2012;**27**

[69] European Commission. Impact Assessment Guidelines (SEC 2005) 791. Brussels [Internet]. 2005. Available from: https://ec.europa.eu/transparency/regdoc/rep/2/2005/EN/SEC-2005-791-2-EN-MAIN-PART-1.PDF [Accessed: 10 January 2020]

[70] Cerreta M, Panaro S. From perceived values to shared values: A multi-stakeholder spatial decision analysis (M-SSDA) for resilient landscapes. Sustainability. 2017;**9**(7):1113

[71] Dochy F, Segers M, Sluijsmans D. The use of self-, peer and co-assessment in higher education: A review. Studies in Higher Education. 1999;**24**(3):331-350

[72] Tessaro F. La valutazione dei processi formativi. Roma: Armando Editore; 1997

[73] Chan KMA et al. Opinion: Why protect nature? Rethinking values and the environment. Proceedings of the National Academy of Sciences. 2016;**113**(6):1462-1465

[74] Kim AJ. Game Thinking: Innovate smarter & Drive Deep Engagement with Design Techniques from Hit Games. 2nd edition. Burlingame: Gamethinking.io; 2018

[75] Coppock P, Ferri G. Serious Urban Games. From play in the city to play for the city. In: Media and the City. Urbanism, Technology and Communication. Newcastle: Cambridge Scholars Publishing; 2013. p. 120-134

[76] Schell J. Jesse Schell: When Games Invade Real Life. In: *DICE Summit*. 2010

[77] Deterding S, Khaled R, Nacke LE, Dixon D. Gamification: Toward a definition. In: CHI 2011 Gamification Workshop Proceedings. Vol. 12. Vancouver BC, Canada; 2011

[78] Viola F. Gamification/Pointsfication/ Serious Games. 2011 [Online]. Available from: http://www.gameifications.com/ gamification-pointsification/

[79] McGonigal J. Reality Is Broken: Why Games Make us Better and How They Can Change the World. New York: Penguin; 2011

[80] Viola F. Fabio Viola speaker con TuoMuseo a FareTurismo. 2015[Online]. Available from: http:// www.gameifications.com/?s=tuomuseo

[81] TuoMuseo. tuomuseo.it. 2018. [Online]. Available from: https://www. tuomuseo.it/

[82] Shin D-H, Kim H-K. A study on user experience based Storydoing operating principles. The Journal of Digital Contents Society. 2015;**16**(3):425-436

[83] Leigh A. GDC 2012: Sid Meier on how to See Games as Sets of Interesting Decisions. 2012. Available from: https://www.gamasutra.com/view/ news/164869/GDC_2012_Sid_Meier_ on_how_to_see_games_as_sets_of_ interesting_decisions.php

[84] Morrison B. Gamasutra Blog. 2013. [Online]. Available from: www. gamasutra.com/blog/bricemorrison

[85] Viola F, Cassone VI. L'arte del coinvolgimento: emozioni e stimoli per cambiare il mondo. Milano: Hoepli editore; 2017

[86] Marczewski A. Gamification–A simple introduction. Tips, advice and thoughts on gamification [Internet]. 2012. Available from: https://cutt.ly/ ltLAJTS [Accessed: 13 January 2020]

[87] Freeman B, Hawkins R. Evoke-developing Skills in Youth to Solve the World's Most Complex Problems: The Social Innovators' Framework. 11th ed. Washington, DC: World Bank Education, Technology and Innovation: SABER-ICT Technical Paper Series; 2016

[88] Moccozet L, Tardy C, Opprecht W, Léonard M. Gamification-based assessment of group work. In: 2013 International Conference on Interactive Collaborative Learning (ICL). 2013. pp. 171-179

[89] Getplayoff.com [Internet]. 2020. Available from: https://www.getplayoff. com/ [Accessed: 12 February 2020]

[90] Federculture. Impresa Cultura. Comunità, territori, sviluppo: 14° Federculture. Roma: Gangemi Editore; 2018

Section 6

Game Theory Applications

Chapter 11

Existence of Open Loop Equilibria for Disturbed Stackelberg Games

T.-P. Azevedo Perdicoúlis, G. Jank and P. Lopes dos Santos

Abstract

In this work, we derive necessary and sufficient conditions for the existence of an hierarchic equilibrium of a disturbed two player linear quadratic game with open loop information structure. A convexity condition guarantees the existence of a unique Stackelberg equilibria; this solution is first obtained in terms of a pair of symmetric Riccati equations and also in terms of a coupled of system of Riccati equations. In this latter case, the obtained equilibrium controls are of feedback type.

Keywords: differential games, linear quadratic, Riccati differential equations, Stackelberg equilibrium, worst-case disturbance

1. Introduction

The study of linear quadratic (LQ) games has been addressed by many authors [1–4]. This type of games is often used as a benchmark to assess the game equilibrium strategies and its respective outcomes. In a disturbed differential game, each player calculates its strategy taking into account a worst-case unknown disturbance. In non-cooperative game theory, the concept of hierarchical or Stackelberg games is very important, since different applications in economics and engineering exist [1, 5]. This is also the case of gas networks where a hierarchy may be assigned to its controllable elements—compressors, sources, reductors, etc... Also, for this application, the modelling as a disturbed game makes a lot of sense, since the unknown offtakes of the network can be modelled as unknown disturbances. Further research on Stackelberg games can be found for instance in AbouKandil and Bertrand [6]; Medanic [7]; Yong [8]; Tolwinski [9].

No assumptions/constraints are made of the disturbance. To be easier to understand the hierarchical concept, we consider only two players. Therefore, we study a LQ game of two players with Open Loop (OL) information structure where the players choose its strategy according to a modified Stackelberg equilibrium. Player-1 is the follower and chooses its strategy after the nomination of the strategy of the leader. Player-2, the leader, chooses its strategy assuming rationality of the follower. Both players find their strategies assuming a worst-case disturbance.

In this work, we consider a finite time horizon, where for applications this is chosen according to the periodicity of the operation of the problem being studied.

The disturbed case of the representation of optimal equilibria for noncooperative games has been studied [10, 11] considering a Nash equilibrium. It is the aim of this paper to generalise the work of Jank and Kun to Stackelberg games and extend the results presented in Freiling and Jank [12]; Freiling et al. [13] to the disturbed case. To calculate the controls, we use a value function approach, appropriately guessed. Thence, we obtain sufficient conditions of existence of these controls and its representation in terms of the solution of certain Riccati equations. Furthermore, a feedback form of the worst-case Stackelberg equilibrium is obtained.

In a future paper, we expect to present analogous conditions using an operator approach.

In Section 2, we define the disturbed LQ game and define Stackelberg worst-case equilibrium. In Section 3, we derive sufficient conditions for the existence of a worst-case Stackelberg equilibrium under OL information structure and investigate how are these solutions related to certain Riccati differential equations. Section 4 concludes the paper and outlines some directions for future work.

2. Fundamental notions

We start with the concept of best reply:

Definition 2.1. (Best reply) Let Γ_N be a N-player differential game. For $i \in \{1, \dots, N\}$,

$$\gamma_{(-i)} := (\gamma_1, \dots, \gamma_{i-1}, \gamma_{i+1}, \dots, \gamma_N) \in \otimes_{j \neq i} \mathcal{U}_j.$$

We say that $\tilde{\gamma}_i$ is the best reply against $\gamma_{(-i)}$ if

$$J_i(\gamma_1, \dots, \gamma_{i-1}, \tilde{\gamma}_i, \gamma_{i+1}, \dots, \gamma_N) \leq J_i(\gamma_1, \dots, \gamma_N)$$

holds for any strategy $\gamma_i \in \mathcal{U}_i$. We denote the set of all best replies by $\mathcal{R}_i\left(\gamma_{(-i)}\right)$.

We study games of quadratic criteria, defined in a finite time horizon $[t_0, t_f] \subset \mathbb{R}$ and subject to a linear dynamics, controlled players and also an unknown disturbance. Hereby also consider $u_j = \gamma_j\left(t, \eta_j\right)$, where η_j is the information structure of Player-j. In this case, $\eta_j, j = 1, \dots, N,$ is of OL type.

Definition 2.2. (Linear Quadratic (LQ) differential game) Let Γ_N be an $N-$ player differential game finite time horizon $T = [t_0, t_f]$. Suppose further that:

i. the dynamics of the game are assumed to obey a linear differential equation

$$\dot{x}(t) = A(t)x(t) + \sum_{j=1}^{N} B_j(t)u_j(t) + C(t)w(t), \qquad (1)$$

$$x(t_0) = x_0.$$

In this equation, $t \in \mathcal{T}$, where the initial t_0 and the final t_f are finite and fixed, the state $x(t)$ is an $n-$ dimension vector of continuous functions defined in $\mathcal{T}$ and with $x\left(t_f\right) = x_f$. The controls $u_i, i = 1, \dots, N,$ are square (Lebesgue) integrable and the m_i- dimension vector of continuous functions is also defined in $\mathcal{T}$. Also, the disturbance $w(t) \in \mathcal{L}^m(\mathcal{T})$. The different matrices are of adequate dimension and with elements continuous in $\mathcal{T}$.

i. the performance criteria are of the form

$$J_i\left(u_i, u_{(-i)}, w\right) = \mathcal{K}\left(x\left(t_f\right)\right) + \int_{t_0}^{t_f} \Psi\left(u_i, u_{(-i)}, w\right) dt. \qquad (2)$$

where

$$\mathcal{K}(x(t_f)) = x^T(t_f)K_{if}x(t_f) \tag{3}$$

$$\begin{aligned}\Psi(u_i, u_{(-i)}, w) &= x^T(t)Q_i(t)x(t) + w^T(t)P_i(t)w(t)\\ &+ \sum_{j=1}^{N} u_j^T(t)R_{ij}(t)u_j(t),\end{aligned} \tag{4}$$

with symmetric matrices $K_{if} \in \mathbb{R}^{n\times n}$ and symmetric, piecewise continuous and bounded matrix valued functions $Q_i(t) \in \mathbb{R}^{n\times n}$, $R_{ij}(t) \in \mathbb{R}^{m_i\times m_j}$ and $P_i(t) \in \mathbb{R}^{m\times m}$, $i = 1, 2, \dots, N$.

We observe that no cost functional is assigned to the disturbance term because no constraints can be applied on an "unpredictable" parameter. In what follows, we consider $N = 2$. To extend the theory to $N > 2$, since this is an hierarchical solution, we need to define the structure of the leaders and followers in the game. We can even have more than two hierachy levels. We assume that Player-2 is the leader and Player-1 is the follower.

The leader seeks a strategy $u_2^*(t)$ in OL information structure and announces it before the game starts. This strategy is found knowing how the follower reacts to his choices. The follower calculates its strategy as a best reply to the strategy announced by the leader.

Problem 2.1. Find the control $u_i^* \in \mathcal{U}_i, i = 1, 2,$ in $\mathcal{T}$ for which $J_i^*\left(u_i, u_{(-i)}^*, w\right), i = 1, 2,$ is minimal when subject to constraints $u_i^*(t) = \gamma_i^*(t, \eta_i(t)), i = 1, 2,$ and (1) and considering a worst-case disturbance.

Consider $\mathcal{U}_i, i = 1, 2,$ the sets of functions such that (1) is solvable and J_i exists, with $u_i, i = 1, 2,$ in these conditions $\mathcal{U}_i, i = 1, 2, \mathcal{W}$ are said the sets of admissible controls and disturbance, respectively.

Definition 2.3. (Stackelberg equilibrium) Let Γ_2 be a 2-person differential game, we define the Stackelberg/worst-case equilibrium in two stages.

1. A function $\hat{w}_i(u) \in \mathcal{W}$ is called the worst-case disturbance, from the point of view of the ith player belonging to the set of admissible controls, if

$$J_i^*\left(u_i, u_{(-i)}^*, \hat{w}_i\right) \geq J_i^*\left(u_i, u_{(-i)}^*, w\right), \quad i = 1, 2, \tag{5}$$

holds for each $w \in \mathcal{W}$. There exists exactly one worst-case disturbance from the point of view of the ith player according to every set of controls.

2. We say that the controls (u_1^*, u_2^*) form a worst-case Stackelberg equilibrium if

i. The leader chooses u_2^* such that

$$\max_{\gamma_1 \in \mathcal{R}_1(u_2^*)} J_2(\gamma_1, u_2^*, \hat{w}_2) \leq \max_{\gamma_1 \in \mathcal{R}_1(u_2)} J_2(\gamma_1, u_2, \hat{w}_2)$$

for all $u_2 \in \mathcal{U}_2$.

ii. The follower then chooses u_1^* such that

$$\mathcal{R}_1(u_2) = \{u_1 | J_1(u_1, u_2, \hat{w}_1) \le J_1(\gamma_1, u_2, \hat{w}_1)\}.$$

To guarantee the uniqueness of OL Stackelberg solutions, matrices are assumed to satisfy $K_{if} \geq 0, Q_i \geq 0, R_{ij} > 0, i \neq j$ and $R_{ii} \geq 0, i, j = 1, \dots, N$ in $\mathcal{T}$ Simaan and Cruz [14].

In what follows, we drop the dependence of the parameters in t to reduce the length of the formulas.

3. Sufficient conditions for the existence of OL Stackelberg equilibria

In this section, we withdraw sufficient conditions for the existence of the worst-case Stackelberg equilibrium, using a value function approach.

A disturbed differential LQ game as defined in Definition 2.2 is said *playable* if there exists a unique Stackelberg worst-case equilibrium.

Theorem 3.1. Let the solution of the Riccati differential equation

$$\begin{aligned} \dot{E}_1 &= -E_1 A - A^T E_1 - Q_1 + E_1(S_1 + T_1)E_1, \\ E_1(t_f) &= K_{1f}, \end{aligned} \tag{6}$$

with $S_1 = B_1 R_{11}^{-1} B_1^T$ and $T_1 = C P_1^{-1} C^T$ exist on $\mathcal{T}$.

For any given admissible OL control of the leader, u_2, define $e_1 \in \mathbb{R}^n, d_1 \in \mathbb{R}$ by

$$\begin{aligned} \dot{e}_1 &= E_1(S_1 + T_1)e_1 - 2E_1 B_2 u_2 - A^T e_1^T, \\ e_1(t_f) &= 0 \end{aligned} \tag{7}$$

$$\begin{aligned} \dot{d}_1 &= -(u_2^T R_{12} + e_1^T B_2) u_2 + \frac{1}{4} e_1^T (S_1 + T_1) e_1, \\ d_1(t_f) &= 0. \end{aligned} \tag{8}$$

Then, the following identity holds:

$$\begin{aligned} 2J_1(u_1, u_2) &= x_0^T E_1(t_0) x_0 + x_0^T e_1(t_0) + d_1(t_0) \\ &\quad + \int_{t_0}^{t_f} \|z_1(t)\|_{R_{11}}^2 dt + \int_{t_0}^{t_f} \|z(t)\|_{P_1}^2 dt, \end{aligned} \tag{9}$$

where $\|z_1\|_{R_{11}}^2 = z_1 R_{11} z_1$ with

$$z_1 = u_1 + R_{11}^{-1} B_1^T \left(E_1 x + \frac{1}{2} e_1 \right)$$

and $\|z\|_{P_1}^2 = z P_1 z$ with

$$z = w + P_1^{-1} C^T \left(E_1 x + \frac{1}{2} e_1 \right)$$

and x a solution of (1).

Proof: The proof is similar to the analogous result for the non-disturbed case Freiling et al. [13].

Theorem 3.2. Let the solution E_1 of (6) exist on $\mathcal{T}$. Then the unique response of the follower to the leader's OL strategy $u_2(t)$ is given by:

$$u_1^* = -R_{11}^{-1}B_1^T\left(E_1x + \frac{1}{2}e_1\right), \tag{10}$$

where the maximum disturbance,

$$w_1^* = -P_1^{-1}C^T\left(E_1x + \frac{1}{2}e_1\right), \tag{11}$$

was considered. E_1 and e_1 are the solutions of (6)–(7) and x is then the solution of

$$\dot{x} = [A - (S_1 + T_1)E_1]x - \frac{1}{2}(S_1 + T_1)e_1 + B_2u_2, \tag{12}$$

$$x(t_0) = x_0. \tag{13}$$

The corresponding minimal costs then are

$$J_{10} = 2J_1(u_1, u_2) = x_0^TE_1(t_0)x_0 + x_0^Te_1(t_0) + d_1(t_0). \tag{14}$$

Proof: We have that the unique OL response of the follower to the leader's announced strategy u_2 (10) under worst-case disturbance (11), that we substitute in the trajectory (1) to obtain:

$$\dot{x} = [A - (S_1 + T_1)E_1]x - \frac{1}{2}(S_1 + T_1)e_1 + B_2u_2.$$

The cost functional minimal value is obtained when we substitute in (9) the minimal control and themaximal disturbance.

Notice that $J_{10}(u_2)$ is not depending on u_1. This, as a matter of fact, is only true if we consider OL information structure, since otherwise u_2 would depend on the trajectory x and hence, via (1), also on u_1. In OL Stackelberg games, the leader tries next to find an optimal OL control u_2 that minimises $J_2(u_1(u_2), u_2)$ while $u_1(u_2)$ is defined by (10).

Theorem 3.3. Let the solution of the Riccati differential Eq. (6) and the solution of

$$\dot{E}_2 = -E_2H - H^TE_2 - Q + E_2(S + T)E_2,$$
$$E_2(t_f) = \begin{pmatrix} K_{2f} & 0 \\ 0 & 0 \end{pmatrix}, \tag{15}$$

with $S_{21} := B_1R_{11}^{-1}R_{21}R_{11}^{-1}B_1^T, S_2 := B_2R_{22}^{-1}B_2^T$ and $T_2 := CP_2^{-1}C^T$. Also $H := \begin{pmatrix} A & -S_1 \\ -Q_1 & E_1T_1 - A^T \end{pmatrix}$, $Q := \begin{pmatrix} Q_2 & 0 \\ 0 & S_{21} \end{pmatrix}$, $S := \begin{pmatrix} S_2 & 0 \\ 0 & 0 \end{pmatrix}$ and $T := \begin{pmatrix} T_2 & T_2E_1 \\ E_1T_2 & E_1T_2E_1 \end{pmatrix}$ exist in $\mathcal{T}$, where $E_2 \in \mathbb{R}^{2n\times 2n}$. Also $B := \begin{pmatrix} B_2 \\ 0_{m_1\times n} \end{pmatrix}$.

For any given control u_2 of the leader, define functions $e_2 \in \mathbb{R}^{3n}$, $v_1, v_w, x \in \mathbb{R}^n$ and $d_2 \in \mathbb{R}$ in $\mathcal{T}$ by the following initial and terminal value problems:

$$\dot{e}_2 = \left(-H^T + E_2(S + T)\right)e_2, \quad e_2(t_f) = 0 \tag{16}$$

$$\dot{d}_2 = \frac{1}{4} e_2^T (S+T) e_2, \quad d_2(t_f) = 0 \tag{17}$$

$$\dot{v}_1 = -Q_1 x + \left(E_1 T_1 - A^T\right) v_1 + E_1 C w, \qquad v_1(t_0) = v_{10} \tag{18}$$

$$\dot{x} = Ax - S_1 v_1 + B_2 u_2 + Cw, \quad x(t_0) = x_0, \tag{19}$$

with $v_1 := \left(E_1 + \frac{1}{2} e_1\right)$.
Then, we obtain

$$u_1^* = -R_{11}^{-1} B_1^T v_1,$$
$$w_1^* = -P_1^{-1} C^T v_1,$$

and the following identity

$$2J_2(u_1^*, u_2^*, w_2^*) = (x_0^T \ v_{10}) E_2(t_0) \begin{pmatrix} x_0 \\ v_{10} \end{pmatrix} + (x_0^T \ v_{10}) e_2(t_0) + d_2(t_0) + \int_{t_0}^{t_f} \|z_2\|_{R_{22}}^2 dt + \int_{t_0}^{t_f} \|z\|_{P_2}^2 dt,$$

where $y = \begin{pmatrix} x \\ v_1 \end{pmatrix}$, $\|z_2\|_{R_{22}}^2 = z_2 R_{22} z_2$ and

$$z_2 = u_2 + \left(R_{22}^{-1} B_2^T \quad 0_{m_1 \times n}\right) \left(E_2 y + \frac{1}{2} e_2\right)$$

and $0_{m_i \times n}, i = 1, 2$ the $m_i \times n$ dimensional zero matrix and $\|z\|_{P_2}^2 = z P_2 z$ and

$$z = w_2 + P_2^{-1} C_1^T \left(E_2 y + \frac{1}{2} e_2\right).$$

Proof: Consider (10): $u_1^* = -R_{11}^{-1} B_1^T \underbrace{\left(E_1 x + \frac{1}{2} e_1\right)}_{:= v_1}$. Then, differentiate v_1 and substitute the derivatives into the obtained expression using (6), (7) and (8). Also, the optimal control u_1^* and disturbance w_1^* in (11). Hence:

$$\dot{v}_1 = -Q_1 x - A^T v_1,$$
$$\dot{x} = Ax - S_1 v_1 + B_2 u_2 + Cw.$$

Hence defining $H := \begin{pmatrix} A & -S_1 \\ -Q_1 & E_1 T_1 - A^T \end{pmatrix}$, $B := \begin{pmatrix} B_2 \\ O_{n \times m_2} \end{pmatrix}$ and $C_1 := \begin{pmatrix} I \\ E_1 \end{pmatrix} C$. We define $y := \begin{pmatrix} x \\ v_1 \end{pmatrix}$ to write these two equations as: (??) as:

$$\dot{y} = Hy + Bu_2 + C_1 w \tag{20}$$

Next, we consider the following value function

$$\tilde{V}_2(t) = V_2(t, y(t)) = y^T E_2 y + e_2^T y + d_2 \tag{21}$$

for some mappings $E_2 : \mathcal{T} \to \mathbb{R}^{2n\times 2n}, e_2 : \mathcal{T} \to \mathbb{R}^{2n}$, and $d_2 : \mathcal{T} \to \mathbb{R}^2$, where E_2 is symmetric for each $t \in T$.

We consider (21), where we substitute (20):

$$\begin{aligned}
\frac{d\tilde{V}_2}{dt} &= \frac{d}{dt}\left(y^T E_2 y + e_2^T y + d_2\right) \\
&= \dot{y}^T E_2 y + y^T \dot{E}_2 y + y^T E_2 \dot{y} + \dot{e}_2^T y + e_2^T \dot{y} + \dot{d}_2 \\
&\quad + x^T Q_2 x + u_1^T R_{21} u_1 + u_2^T R_{22} u_2 + w^T P_2 w - \psi_2 \\
&= \left(y^T H^T + u_2^T B^T w^T C_1^T\right) E_2 y \\
&\quad + y^T \dot{E}_2 y + y^T E_2 (Hy + Bu_2 + C_1 w) \\
&\quad + \dot{e}_2^T y + e_2^T (Hy + Bu_2 + C_1 w) + \dot{d}_2 \\
&\quad + y^T \underbrace{\begin{pmatrix} Q_2 & 0 \\ 0 & S_{21} \end{pmatrix}}_{:=Q} y + u_2^T R_{22} u_2 + w^T P_2 w - \psi_2
\end{aligned}$$

Now we associate certain terms

$$\begin{aligned}
&= y^T \left(H^T E_2 + \dot{E}_2 + E_2 H + Q\right) y \\
&\quad + (u_2 - y_2)^T R_{22} (u_2 - y_2) \\
&\quad + y_2^T R_{22} u_2 + u_2^T R_{22} y_2 - y_2^T R_{22} y_2 \\
&\quad + (w - \alpha)^T P_2 (w - \alpha) \\
&\quad + w^T P_2 \alpha + \alpha^T P_2 w - \alpha^T P_2 \alpha \\
&\quad + u_2^T \left(B^T E_2 y + B^T e_2\right) + \left(y^T E_2 B + \frac{1}{2} e_2^T B\right) u_2 \\
&\quad + w^T \left(C_1^T E_2 y + C_1^T e_2\right) + \left(y^T E_2 C_1 + \frac{1}{2} e_2^T C_1\right) w \\
&\quad (\dot{e}_2^T + e_2^T H) y + \dot{d}_2 - \psi_2
\end{aligned}$$

and furthermore

$$\begin{aligned}
&= y^T \left(H^T E_2 + \dot{E}_2 + E_2 H + Q\right) y - \psi_2 \\
&\quad + (u_2 - y_2)^T R_{22} (u_2 - y_2) - y_2^T R_{22} y_2 \\
&\quad + (w - \alpha)^T P_2 (w - \alpha) - \alpha^T P_2 \alpha \\
&\quad + u_2^T \left(B^T E_2 y + B^T \frac{1}{2} e_2 + R_{22} y_2\right) \\
&\quad + \left(y^T E_2 B + \frac{1}{2} e_2^T B y_2^T R_{22}\right) u_2 \\
&\quad + w^T \left(C_1^T E_2 y + C_1^T \frac{1}{2} e_2 + P_2 \alpha\right) \\
&\quad + \left(y^T E_2 C_1 + \frac{1}{2} e_2^T C_1 - \alpha^T P_2\right) w \\
&\quad (\dot{e}_2^T + e_2^T H) y + \dot{d}_2 - \psi_2
\end{aligned}$$

Consider

$$R_{22}y_2 + \begin{pmatrix} B_2^T & O_{m_2\times n} \end{pmatrix}\left(E_2y + \frac{1}{2}e_2\right) = 0$$

and also

$$C_1^T\left(E_2y + \frac{1}{2}e_2\right) + P_2\alpha = 0$$

If $R_{22} > 0$ then $y_2 = -R_{22}^{-1}\begin{pmatrix} B_2^T & O_{m_2\times n} \end{pmatrix}\left(E_2y + \frac{1}{2}e_2\right)$. If $P_2 > 0$ then $\alpha = -P_2^{-1}C_1\left(E_2y + \frac{1}{2}e_2\right)$.

Define $S := \begin{pmatrix} S_2 & 0 \\ 0 & 0 \end{pmatrix}$ and $T := \begin{pmatrix} T_2 & T_2E_1 \\ E_1T_2 & E_1T_2E_1 \end{pmatrix}$. Substitute this y_2 and α in the calculations:

$$\begin{aligned} &= y^T\left(H^TE_2 + \dot{E}_2 + E_2H + Q - E_2(S+T)E_2\right)y \\ &+ (u_2 - y_2)^T R_{22}(u_2 - y_2) + (w-\alpha)^T P_2(w-\alpha) \\ &+ \left(\dot{e}_2^T + e_2^TH - E_2(S+T)\right)y \\ &+ \dot{d}_2 - \frac{1}{4}e_2^T(S+T)e_2 - \psi_2 \end{aligned}$$

Considering:

$$\begin{aligned} H^TE_2 + \dot{E}_2 + E_2H + Q - E_2(S+T)E_2 &= 0 \\ \dot{e}_2^T + e_2^TH - e_2^T(S+T)E_2 &= 0 \\ \dot{d}_2 - \frac{1}{4}e_2^T(S+T)e_2 &= 0 \end{aligned}$$

that is

$$\begin{aligned} \dot{E}_2 &= -H^TE_2 - E_2H - Q + E_2(S+T)E_2 \\ \dot{e}_2 &= \left(-H^T + E_2(S+T)\right)e_2 \\ \dot{d}_2 &= \frac{1}{4}e_2^T(S+T)e_2 \end{aligned}$$

We end up with

$$\begin{aligned} \frac{d\tilde{V}_2(t)}{dt} &= (u_2 - y_2)^T R_{22}(u_2 - y_2) \\ &+ (w-\alpha)^T P_2(w-\alpha) - \psi_2 \end{aligned} \tag{22}$$

Integrating yields:

$$\begin{aligned} \tilde{V}_2(t_f) - \tilde{V}(t) = \int_t^{t_f} &\Big[(u_2 - y_2)^T R_{22}(u_2 - y_2) \\ &+ (w-\alpha)^T P_2(w-\alpha)s\Big]d\tau - \int_t^{t_f}\psi_2 d\tau. \end{aligned}$$

Further, we assume the mappings E_2, e_2, d_2 to be chosen in such a way that the following terminal values hold:

$$E_2(t_f) = K_{2_f}$$
$$e_2(t_f) = 0$$
$$d_2(t_f) = 0$$

Then, we obtain $\tilde{V}_2(t_f) = y^T(t_f)K_{y_f}y(t_f)$ and substituting:

$$\tilde{V}_2(t) = y^T(t_f)K_{y_f}y(t_f) - \int_t^{t_f} \Big[(u_2 - y_2)^T R_{22}(u_2 - y_2) + (w - \alpha)^T P_2(w - \alpha)\Big] d\tau + \int_t^{t_f} \psi_2 d\tau \tag{23}$$

Observe that the rhs of (23) does not depend of $u|_{[t_0,t]}$ and the rls of (23) does not depend of $u_2|_{[t,t_f]}$ Then considering now the infimal value, we recall that:

$$V_2(t,y) = \inf_{u_2|_{[t,t_f]}} \int_t^{t_f} \psi_2(\tau, \hat{y}(\tau), u(\tau)) d\tau + y^T(t_f)K_{2_f}y(t_f)$$

Now, we substitute this into (23) and consider the infimal values over all possible control functions in $[t, t_f]$:

$$\tilde{V}_2(t) = \underbrace{\inf y^T(t_f)K_{2_f}y(t_f) + \int_t^{t_f} \psi_2 d\tau}_{V_2(t,y)} - \inf \int_t^{t_f} \Big[(u_2 - y_2)^T R_{22}(u_2 - y_2) + (w - \alpha)^T P_2(w - \alpha)\Big] d\tau$$

then we have:

$$V_2(t,y) = \tilde{V}(t) + \inf_{u|_{t,t_f}} \int_t^{t_f} \Big[(u_2 - y_2)^T R_{22}(u_2 - y_2) + (w - \alpha)^T P_2(w - \alpha)\Big] d\tau$$

$V_2(t,y)$ equals $\tilde{V}_2(t)$ if $u_2 - y_2 \equiv 0 \forall t \in \mathcal{T}$ and $w - \alpha = 0$. As the leader chooses his strategy assuming rationality of the follower and worst-case disturbance, the follower should take also the worst-case disturbance into account.

To conclude, consider $t = t_0$ and hence:

$$V_2(t_0,y) = \tilde{V}_2(t_0) + \inf_{u|_{t_0,t_f}} \int_{t_0}^{t_f} \Big[(u_2 - y_2)^T R_{22}(u_2 - y_2) + (w - \alpha)^T P_2(w - \alpha)\Big] d\tau$$

Then from (21):

$$V_2(t_0,y) = y_0^T E_2(t_0) y_0 + e_2^T(t_0) y_0 + d_2(t_0) + \inf_{u|_{t_0,t_f}} \int_{t_0}^{t_f} \Big[(u_2 - y_2)^T R_{22} (u_2 - y_2) + (w-\alpha)^T P_2 (w-\alpha) \Big] d\tau$$

Defining $z_2 := u_2 - y_2 = u_2 + \left(R_{22}^{-1} B_2^T \quad 0 \right)\left(E_2 y + \frac{1}{2} e_2 \right)$ and $z := w - \alpha = w + P_2^{-1} C_1 \left(E_2 y + \frac{1}{2} e_2 \right)$, we have:

$$V_2(t_0,y) = y_0^T E_2(t_0) y_0 + e_2^T(t_0) y_0 + d_2(t_0) + \int_{t_0}^{t_f} \|z_2\|_{R_{22}}^2 + \|z\|_{P_2}^2 dt$$

Now, we substitute $y_0 = \begin{pmatrix} x_0 \\ v_{10} \end{pmatrix}$.

The leader may choose its best answer either by accounting directly for its worst-case disturbance or by considering that the follower knows that there is a worst-case disturbance. In this work, the leader takes the worst-case disturbance directly into account.

Notice that in the term

$$J_{20} = \left(x_0^T \quad v_{10} \right) E_2(t_0) \begin{pmatrix} x_0 \\ v_{10} \end{pmatrix}, \tag{24}$$

$x_0, E_2(t_0)$, do not depend on the choice of u_1, u_2. Since we shall study the situation for Player-2 when Player-1 applies his optimal response control defined in (10), we have to set $v_1 = E_1 x + \frac{1}{2} e_1$. From (7), we can see that $v_1(t_0) = v_{10}$ depends on $e_1(t_0)$ and hence also on u_2.

In order to derive from Theorems (3.1) and (3.3) sufficient conditions for the existence of a unique worst-case Stackelberg equilibrium, we must get rid of the u_2-dependence on v_{10}. Therefore, we propose to restrict the set of admissible controls to functions representable in linear feedback form. This is what we do next.

Theorem 3.4. Let the solutions $E_1(t) \in \mathbb{R}^{n \times n}, E_2 \in \mathbb{R}^{2n \times 2n}$ of (6) and (15) exist in $\mathcal{T}$, respectively. Let further the coupled system of equations

$$\dot{K}_1 = -Q_1 - K_1 A - A^T K_1 + K_1 (S_1 + T_1) K_1 + K_1 S_2 K_2, \tag{25}$$

$$\dot{K}_2 = -Q_2 - K_2 A - A^T K_2 + Q_1 p + K_2 S_1 K_1 + K_2 (S_2 + T_2) K_2 + K_2 T_2 E_1 p, \tag{26}$$

$$\dot{p} = -pA - S_{21} K_1 + S_1 K_2 + (A - T_1 E_1) p + p S_1 K_1 + p (S_2 + T_2) K_2 + p T_2 E_1 p, \tag{27}$$

admits a solution in $\mathcal{T}$.

Then, there exists a unique open loop disturbed Stackelberg equilibrium in feedback synthesis which is given by

$$u_1^*(t) = -R_{11}^{-1}(t) B_1^T(t) K_1(t) x(t), \tag{28}$$

$$u_2^*(t) = -R_{22}^{-1}(t) B_2^T(t) K_2(t) x(t), \tag{29}$$

considering worst-case disturbamces w_i^* and where $x(t)$ is a solution of the closed loop equation

$$\begin{aligned} \dot{x} &= [A - S_1K_1 - (S_2 + T_2)K_2 - T_2E_1p]x, \\ x(t_0) &= x_0. \end{aligned} \tag{30}$$

The minimal cost for the follower, $J_{10}(u_2^*)$, is as in (14), and for the leader is

$$J_{20}(u_1^*, u_2^*) = \frac{1}{2}\left[x_0^T(I_n, K_1^T(t_0))E_2(t_0)(t_0)\begin{pmatrix} I_n \\ K_1(t_0)\end{pmatrix}x_0 + e_2^T(t_0)\begin{pmatrix} I_n \\ K_1(t_0)\end{pmatrix}x_0 + d_2(t_0)\right]$$

where $e_2(t_0), d_2(t_0)$ are determined by (16) and (17), respectively.

Proof: The proof is similar to the analogous result for the non-disturbed case [13].

From the convexity assumptions, it follows that S_1, S, Q_1, Q and $E_1(t_f), E_2(t_f)$ are all semidefinite. Therefore, as far as the convexity conditions hold, the standard Riccati matrix Eqs. (6) and (15) are globally solvable in $(-\infty, t_f]$ [15].

It still remains the following questions to be answered (i) direct criteria for solvability of these equations if the convexity assumption is guaranteed as well as (ii) solvability of the coupled system of Eqs. (25)–(27).

Actually, this system of equations can also be written as a single, nonsymmetric Riccati matrix differential equation. Hence:

$$\begin{aligned} \begin{pmatrix} \dot{K}_1 \\ \dot{K}_2 \\ \dot{p} \end{pmatrix} &= \begin{pmatrix} Q_1 \\ Q_2 \\ 0 \end{pmatrix} - \begin{pmatrix} K_1 \\ K_2 \\ p \end{pmatrix} A \\ &+ \begin{pmatrix} -A^T & 0 & 0 \\ 0 & -A^T & Q_1 \\ -S_{21} & S_1 & A - T_1E_1 \end{pmatrix}\begin{pmatrix} K_1 \\ K_2 \\ p \end{pmatrix} \\ &+ \begin{pmatrix} K_1 \\ K_2 \\ p \end{pmatrix}(S_1 + T_1, \quad S_2, \quad 0)\begin{pmatrix} K_1 \\ K_2 \\ p \end{pmatrix} \\ \begin{pmatrix} K_1 \\ K_2 \\ p \end{pmatrix}(t_f) &= \begin{pmatrix} K_{1f} \\ K_{2f} \\ 0 \end{pmatrix}. \end{aligned} \tag{31}$$

As it can be easily observed, all these Riccati equations are of nonsymmetric type:

$$\begin{aligned} \dot{W} &= B_{21} - WB_{11} + B_{22}W + WB_{12}W, \\ W(t_f) &= W_f, \end{aligned} \tag{32}$$

where W is a matrix of order $k \times n$ whose coefficients are of adequate size. See AbouKandil et al. [16] for results on the existence of solution of Riccati equations.

4. Discussion and conclusions

High dimension problems appeal to the use of hierarchic and decentralised models as differential games. One example of these problems is large networks, as for instance the management and control of high pressure gas networks. Since this is a large dimension and geographically dispersed problem, a decentralised formulation captures the non-cooperative nature, and sometimes even antagonistic, of the different stake-holders in the network.

The network controllable elements can be seen as players that seek their best settings and then interact among themselves to check for network feasibility. The equilibrium sought by the players depends on the way the players are organised among themselves. It makes some sense to have some autonomous elements that run the network and others follow, as is the case of a main inlet point of a country, as it happens with the inlet of Sines in the portuguese network. The ultimate goal of the network is to meet customers' demand at the lowest cost. As the main variation of the problem is due to the off-takes, these may be seen as perturbations to nominal consumption levels of a deterministic model.

Therefore, it makes some sense to view the gas transportation and distribution system as a disturbed Stackelberg game where the players play against a worst-case disturbance, that means a sudden change in weather conditions from one period of operation to the other. Neverthless, the theory is not ready, and also having in mind the development of algorithms, direct solution methods, and explicit solution representations need to be further investigated. In this work, we have obtained sufficient conditions for the existence of the solution of a 2-player game. However, direct criteria for solvability of this problem needs more work. Also, the solvability of the coupled system of Eqs. (25)–(27) has to be further investigated. Also, we would like to solve the same problem using an operator approach.

Similarly to what we have done in the past for Nash games, we would like to study this problem considering the underlying dynamics as a repetitive process, that seems to be adequate to capture the behaviour seemingly periodic of the network. Also, the boundary control of the network depends on the type of strategy sought by the players. The structure of these versions of the problems need to be examined.

The obtained results, in every stage of the work, should be applied to a single pipe and ideally using some operational data. Furthermore, we expect to apply the work to a simple network, which is not exactly a straightforward extension.

Acknowledgements

I would like to thank the reviewer for his valuable suggestions.

This work has been financed by National Funds through the Portuguese funding agency, FCT – Fundao para a Cincia e a Tecnologia under project: (i) UID/EEA/00048/2019 for the first author and (ii) UID/EEA/50014/2019 for the third author.

Author details

T.-P. Azevedo Perdicoúlis[1*], G. Jank[2] and P. Lopes dos Santos[3]

1 ISR, Engineering Department at ECT UTAD, University of Coimbra, Portugal

2 RWTH-Aachen, Germany

3 INESC TEC and EE Department at FEUP, University of Porto, Portugal

*Address all correspondence to: tazevedo@utad.pt

References

[1] Bagchi A. Differential games in economic models. In: Lecture Notes in Control and Information Sciences. Vol. 64. Springer Verlag; 1984

[2] Eisele T. Nonexistence and nonuniqueness of open-loop equilibria in linear quadratic differential games. Journal of Optimization Theory and Applications. 1982;**37**:443-468

[3] Lukes DL, Russel DL. A global theory for linear quadratic differential games. Journal of Mathematical Analysis and Applications. 1971;**33**:96-123

[4] Mehlmann A. Applied Differential Games. New York: Springer Verlag; 1988

[5] Basar T, Olsder G. Dynamic noncooperative game theory. 2nd ed. London: SIAM; 1999

[6] Abou-Kandil H, Bertrand P. Analytical solution for an open-loop Stackelberg game. IEEE Transactions on Automatic Control. 1985;**AC-30**: 1222-1224

[7] Medanic J, Radojevic D. Multilevel Stackelberg strategies in linear quadratic systems. Journal of Optimization Theory and Applications. 1987;**24**: 485-479

[8] Pan L, Yong J. A differential game with multi-level of hierarchy. Journal of Mathematical Analysis and Applications. 1991;**161**:522-544

[9] Tolwinski B. A Stackelberg solution of dynamic games. IEEE Transactions on Automatic Control. 1973;**AC-28**: 85-93

[10] Jank G, Kun G. Optimal control of disturbed linear quadratic differential games. European Journal of Control. 2002;**8**(2):152-162

[11] Kun G. Stabilizability, controllability and optimal strategies of linear and nonlinear dynamical games [Ph.D. Thesis]. RWTH Aachen; 2000

[12] Freiling G, Jank G. Existence of open-loop Stackelberg equilibria. In: Proceedings of the Mathematical Theory of Networks and Systems, MTNS 98. Symposium held at Padova, Italy; 1998

[13] Freiling G, Jank G, Lee S-R. Open-loop Stackelberg equilibria in linear quadratic differential games. Journal of Optimization Theory and Applications. 2001;**110**(3):515-544

[14] Simaan M, Cruz JB. On the Stackelberg strategy in nonzero-sum games. Journal of Optimization Theory and Applications. 1973;**11**(5):533-555

[15] Knobloch HW, Kwakernaak H. Lineare KontrollTheorie. London: Springer Verlag; 1985

[16] Abou-Kandil H, Freiling G, Ionescu V, Jank G. Matrix Riccati Equations in Control and Systems Theory. Basel: Birkhäuser Verlag; 2003

Chapter 12

Nash Equilibrium Study for Distributed Mode Selection and Power Control in D2D Communications

Sameh Najeh and Ammar Bouallegue

Abstract

One of the main challenges of LTE-advanced (LTE-A) is to recover the local-area services and improve spectrum effciency. In order to reach those goals technical capabilities are required. D2D is a promising techniques for the 5G wireless communications system using several applications, as: network traffic offoading, public safety, social services and applications such as gaming and military applications. In this chapter, we investigate both mode selection and distributed power control in D2D system. Indeed, the mode selection is provided while respecting a predetermined SINR threshold relative to cellular and D2D users. The amount of minimum and maximum power are then derived to fulffill the predetermined requirements, by limiting the interference created by underlaid D2D users. In order to realize our proposed power control step, a new distributed control approach is proposed using game theory tools for several cellular and D2D users. This distributed approach is based on the mode selection strategy already proposed in the previous step. Finally, simulations were established in order to compare the proposed distributed algorithm in terms of coverage probability which is based on game theory, with other conventional centralized algorithms.

Keywords: mode selection, power control, distributed, Nash equilibrium

1. Introduction

The Internet of Things (IoT) is a developing and promising innovation, which were able to revolutionize the world [1]. IoT manages low-powered gadgets, using the internet by interacting with one another. IoT interconnect "Things" and also helps in machine-to-machine (M2M) communication, which is a way of data communication between varied gadgets without human intercession [2].

IoT applications can be classified into six main categories, such as [1]: smart cities, smart business, smart homes, healthcare, security and surveillance. Regarding these different applications, several requirements should be maintained, like [2]: (1) high scalability, (2) security and privacy, (3) high capacity, (4) security and privacy, (5) energy saving, (6) reduced latency, (7) quality of service (QoS), (8) built-in redundancy, (9) heterogeneity and (10) efficient network and spectrum.

The 5G enabled IoT contains a number of key communication techniques for IoT applications, in order to make the network with faster speeds and greater accessibility. A network that reaches all over the world and is accessible to all. Since 5G technology offers greater connectivity, more and more applications for this technology are likely to come to the field. Four main categories can be cited in the this context: (1) Wireless Network Function Virtualization, (2) Architecture of 5–IoT, (3) Heterogeneous Network (HetNet) and (4) Device to Device (D2D) Communication.

In this chapter, we mainly focus on the last type especially on D2D Communication [3–8], which allows the exchange of data between user equipment without the use of the base station. The short distance communication between two devices (D2D) becomes a challenging way to transmit data, since it benefits the 5–IoT with low power consumption, load balancing and better QoS for edge users. Indeed, in IoT over 60% of applications require low power, a long battery and also wide connectivity coverage. Hence, for these reasons more light should needs to be shed on low-power wireless networks and their prospects in meeting these requirements. Integrating D2D in cellular networks poses challenges and design problems, in order to offer adequate Radio Resource Management (RRM) schemes [4–6, 9–11] and this taking into account all the constraints imposed by the different users. As has already been mentioned in the literature, RRM techniques can be classified into four groups as: (1) Mode Selection (MS): where the Mobile Station determines whether D2D candidates in the proximity of each other should communicate in direct mode using the D2D link or in cellular mode [3–5], (2) Power Control (PC): is an efficient solution to mitigate the interference for D2D underlaid cellular network, in order to improve the overall of the system [6, 9]. (3) Pairing: is a concept which exists only when D2D links are reusing cellular resources and consists on assigning one cellular uplink user (CUE) and one or more D2D uplink user (DUE) links for each resource block [18] and (4) Resource Allocation: is a process of selecting radio resources for each cellular and D2D link, this can be done jointly with MS and pairing [3, 4].

Several approaches have already been proposed in the literature in order to achieve MS and PC management, these approaches can be: (1) Centralized management: where the base station (BS) allocates directly to the designated DUE and require the knowledge of D2D links' Channel State Information (CSI) at the BS level and (2) Distributed (or decentralized) management, in which the BS informs D2D users which Resource Blocks (RBs) they can use. In this chapter, we focus on the RRM algorithms in underlay D2D communication, for both centralized and distributed MS and PC, in order to improve the overall of the system using game theory tools. In [3–7, 9, 11, 13–17, 19–22], the authors have proposed centralized and distributed PC approaches with perfect channel state information (CSI) using powerful mathematical tools, such as game theory [7–9], stochastic approximation [20], mechanism design [21] etc.

In fact, Game theory (GT) is a branch of applied mathematics that provides models and tools for analyzing situations where multiple rational users interact to achieve their goals [23, 24]. Several examples based on Wireless Communications are investigated in the literature, as in PC, congestion control, load balancing, etc. In [6], a centralized and distributed PC algorithms are developed and evaluated for a D2D underlaid cellular system using stochastic geometry. The authors in [9] have focused on maximizing the total sum-rate in an heterogeneous network (HetNet) via game theoretic approaches. The authors in [11] have proposed a distributed PC, based on an appropriate interference management scheme in D2D underlaid Cellular Network by using GT approach. An iterative distributed power allocation algorithm for the two kinds of game: pure and mixed has investigated. Distributed vs. centralized MS and PC approaches have been suggested in [25] using GT tools.

This chapter investigates both MS and PC in D2D communications using centralized and distributed approaches based on GT tools. In the proposed MS approach, the CUE and DUE list, the minimum and maximum quantities of power are derived according to CUE and DUE signal-to-interference plus-noise-ratio (SINR) thresholds. The expression of the minimum amount of power known in the literature as the Pareto power [6, 9, 11], has always been used until now without mathematical proof. We propose in this chapter to show mathematically this Pareto power, which is considered as a key of the PC process. Then, a pure strategy non-cooperative game between the two kind of users is modeled as a distributed PC approach, based on the derived minimum and maximum power, where two utility functions are investigated for both type of users. This chapter reviews the work previously published by the authors in [12]. For this, all the proofs and demonstrations of the different results stated in this chapter will not be provided since they are already explained in [12].

The structure of this chapter is given as: In Section 2, the system model of a D2D communication and CUE and DUE SINR and coverage probability expressions are defined. In Section 3, the closed form expression of both minimum and maximum amounts of power with a mathematical demonstration are provided, based on the predetermined CUE and DUE SINR thresholds. In Section 4, our proposed centralized MS and PC approaches are investigated, which consists of generalizing a classical centralized MS and PC approaches. The Section 5, outlines an iterative NE distributed power approach which is proposed for both CUE and DUE and is based on the minimum and maximum amounts of power, derived from Section 3 and on the GT tools. This proposed distributed approach aims to achieve a better compromise between different users in terms of allocated powers is presented in Section 6. Several simulations are provided in order to assess the performance of the allocation approaches of the MS and PC thus proposed in Section 7. Section 8 is followed with conclusion and future scope.

2. System model

In this section, a D2D uplink underlaid cellular network is considered illustrated in this section, which is shown in **Figure 1**. A system which is composed by a single-cell cellular network, where K_1 CUE and K_2 DUE communicate with as it is illustrated in **Figure 1**, where the BS is in the center of circular coverage area and each D2D user refers to a source-destination pair. The total number of users K is defined as $K = K_1 + K_2$. We assume that all users (CUE and DUE) are drawn in a circular disk C with radius R and are randomly distributed in the whole $\mathbb{R}^2$ and modeled as an independently homogeneous Poisson Point Process (PPP) Φ with density λ. For each kind of user (CUE or DUE) k, we denote y_k as the received signal defined as follows.

- CUE mode: if $1 \le k \le K_1$, y_k is the received signal at the CUE k from the BS.
- DUE mode: if $K_1 + 1 \le k \le K$, y_k is the received signal at the kth DUE receiver from the kth DUE transmitter.

Let $g_{k,i}$ denotes the instantaneous channel gain from the kth transmitter to the ith receiver, where $k, i \in \mathbf{K} = \{1, ..K\}$. Further, we denote $\mathbf{K_1} = \{1, ..K_1\}$ and $\mathbf{K_2} = \{K_1 + 1, ..K\}$.

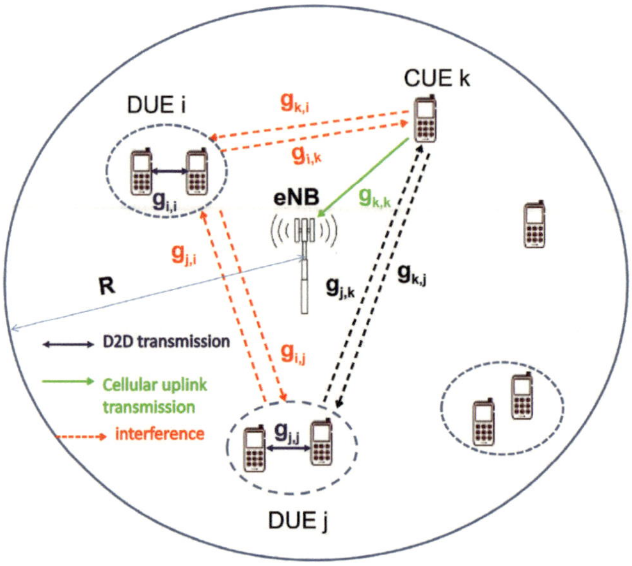

Figure 1.
System model of D2D communication.

2.1 CUE and DUE SINR expressions

In order to ensure a QoS in terms of γ_c^{th} and γ_d^{th}, as SINR thresholds of both CUE and DUE (respectively), we assume the following statement for each user k, as performance yardsticks

$$\begin{cases} \gamma_k(\mathbf{P}) = \dfrac{g_{k,k}p_k}{\sum_{i=1,i\neq k}^{K} g_{k,i}p_i + \sigma^2} \geq \gamma_c^{th}, \forall k \in \mathbf{K_1} : \text{cellular mode}, \quad (1.a) \\ \gamma_k(\mathbf{P}) = \dfrac{g_{k,k}p_k}{\sum_{i=1,i\neq k}^{K} g_{k,i}p_i + \sigma^2} \geq \gamma_d^{th}, \forall k \in \mathbf{K_2} : \text{D2D mode}, \quad (11.b) \end{cases} \tag{1}$$

where p_k is the amount of the transmit powers for the kth user (CUE or DUE), σ^2 is the receiver noise power, $\mathbf{P} = (p_1, ..p_K)$ is the vector of transmit powers and $g_{k,i}$ is defined as follows

$$g_{k,i} = |h_{k,i}|^2 d_{k,i}^{-\alpha}, \tag{2}$$

where, $h_{k,i}$ and $d_{k,i}$ are respectively the distance-independent fading and the distance from the transmitter k to receiver i and α is the path loss.

Let us define for each user k (CUE and DUE), the SINR threshold γ_k^{th}, as

$$\gamma_k^{th} = \begin{pmatrix} \gamma_c^{th} \text{ if } k \in \mathbf{K_1} : \text{Cellular mode} \\ \gamma_d^{th} \text{ if } k \in \mathbf{K_2} : \text{D2D mode}. \end{pmatrix} \tag{3}$$

2.1.1 CUE and DUE coverage probabilities

In order to simplify the notations used in the chapter, we will consider vector rather than analytical expressions. According to each kind of user, we define the coverage probabilities expressions denoted as $P_{c,cov}\left(\mathbf{P},\Gamma_c^{th}\right)$ and $P_{d,cov}\left(\mathbf{P},\Gamma_d^{th}\right)$ for both CUE and DUE (respectively) as

$$P_{c,cov}\left(\mathbf{P},\Gamma_c^{th}\right) \triangleq Prob\left(\Gamma_c(\mathbf{P}) \geq \Gamma_c^{th}\right), \tag{4}$$

$$P_{d,cov}\left(\mathbf{P},\Gamma_d^{th}\right) \triangleq Prob\left(\Gamma_d(\mathbf{P}) \geq \Gamma_d^{th}\right), \tag{5}$$

where,

$$\Gamma_c(\mathbf{P}) = \left(\gamma_1(\mathbf{P}),..\gamma_{K_1}(\mathbf{P})\right), \quad \Gamma_c^{th} = \left(\gamma_c^{th},..\gamma_c^{th}\right). \tag{6}$$

$$\Gamma_d(\mathbf{P}) = \left(\gamma_{K_1+1}(\mathbf{P}),..\gamma_K(\mathbf{P})\right), \quad \Gamma_d^{th} = \left(\gamma_d^{th},..\gamma_d^{th}\right). \tag{7}$$

3. The minimum and maximum amount of power: $(\mathbf{P}_{\mathbf{min}}, \mathbf{P}_{\mathbf{max}})$

This section investigates the study of the existence of the two minimum and maximum powers $\mathbf{P}_{\mathbf{min}}$ and $\mathbf{P}_{\mathbf{max}}$, necessary to verify the constraints imposed by the previous system (1). First, based on this system (1), the minimum power $\mathbf{P}_{\mathbf{min}}$ is derived, already known in the literature under the name of Pareto Power. Second, by limiting the quantity of power by a quantity, which we denote $\mathbf{P}_{\mathbf{max}}$ from $\mathbf{P}_{\mathbf{min}}$, we make sure more that the system (1) remains satisfied as long as we are in the power range $[\mathbf{P}_{\mathbf{min}}, \mathbf{P}_{\mathbf{max}}]$.

To do this, we start by providing another vector form of the (1) system to build this Pareto power.

3.1 Vector form of system

Let us take into consideration the following definition and proposition (1),

Definition 1. If $\mathbf{A} = \left(a_{i,j}\right)_{1\leq i,j\leq K}$ and $\mathbf{B} = \left(b_{i,j}\right)_{1\leq i,j\leq K}$ are two matrices, then we define

$$\mathbf{A} \geq \mathbf{B} \Leftrightarrow a_{i,j} \geq b_{i,j}, \quad 1 \leq i,j \leq K. \tag{8}$$

Proposition 1. The previous Eq. (1) can be written compactly in the following vector form as

$$(\mathbf{I}_K - \mathbf{F})\mathbf{P} \geq \mathbf{b}, \tag{9}$$

where $\mathbf{I}_K$ denotes the identity matrix of order K, $\mathbf{F} = \left(f_{k,i}\right)_{k,i\in\mathbf{K}}$ and $\mathbf{b} = (b_k)_{k\in\mathbf{K}}$, $\forall k, i \in \mathbf{K}$, are defined as below [11, 12].

$$f_{k,i} = \begin{cases} \frac{g_{k,i}}{g_{k,k}}\gamma_k^{th} & \text{if } k \neq i \\ 0 & \text{otherwise} \end{cases} \quad \text{and} \quad b_k = \frac{\sigma^2\gamma_k^{th}}{g_{k,k}}. \tag{10}$$

3.2 Power lower band $\mathbf{P}_{\mathbf{min}}$

All previous works [6, 8, 9, 11] have dealt with the resolution of the problem presented in (9) by using a minimum power $\mathbf{P}_{\mathbf{min}}$ without proof, which is defined as

$$\mathbf{P}_{\mathbf{min}} = (\mathbf{I}_\mathbf{K} - \mathbf{F})^{-1}\mathbf{b}, \tag{11}$$

The authors in [6, 8, 11] have shown that: if $\rho(\mathbf{F}) < 1$ then the matrix $(\mathbf{I}_\mathbf{K} - \mathbf{F})$ is invertible. Next, we propose to suggest another sufficient condition on the matrix $\mathbf{F}$, which guarantees the existence of the inverse of the matrix $(\mathbf{I_K} - \mathbf{F})$, in order to derive the quantity of minimum power $\mathbf{P_{min}}$, already given in Eq. (11).

To do this, we propose theorems, definitions and propositions in order to outline all the necessary steps which allow to build this sufficient condition. Obviously, to make reading easier, all the demonstrations relating to these theorems and propositions are already detailed in [11, 12].

Theorem 1. *We assume that $\rho(\mathbf{F}) < 1$, the following statement is true*

$$(\mathbf{I_K} - \mathbf{F})\mathbf{P} \geq \mathbf{b} \Rightarrow \mathbf{P} \geq \mathbf{P_{min}} = (\mathbf{I_K} - \mathbf{F})^{-1}\mathbf{b}. \tag{12}$$

Hence, if $\rho(\mathbf{F}) < 1$ then the minimum power $\mathbf{P}_{min}$ defined in Eq. (11) exists and we can consider the following notation

$$\mathbf{P_{min}} = \left(\underbrace{P_{min}(1), ..P_{min}(K_1)}_{\mathbf{P_{c,min}}}, \underbrace{P_{min}(K_1+1), \ldots P_{min}(K)}_{\mathbf{P_{d,min}}} \right). \tag{13}$$

We can note in another way

$$\mathbf{P_{min}} = (\mathbf{P_{c,min}}, \mathbf{P_{d,min}}). \tag{14}$$

Proposition 2. Let consider the following iterative power process, relative to each iteration i, as [26].

$$\mathbf{P}(i+1) = \mathbf{F}\mathbf{P}(i) + b. \tag{15}$$

Hence,

$$\lim_{i \to +\infty} \mathbf{F}^i = 0 \Rightarrow \sum_{i=0}^{+\infty} \mathbf{F}^i\mathbf{b} = (\mathbf{I_K} - \mathbf{F})^{-1}\mathbf{b}. \tag{16}$$

3.3 Power upper band $\mathbf{P_{max}}$

To let $\mathbf{P_{max}}$ greater than $\mathbf{P_{min}}$ for all users, it is proposed in this paragraph to build a quantity of power $\mathbf{P_{max}}$ from $\mathbf{P_{min}}$, in order to guarantee the conditions required by the users already depicted in Eq. (11). The two maximum power quantities dedicated to the different CUE and DUE users, denoted as $\mathbf{P}_{c,max}$ and $\mathbf{P}_{d,max}$ (respectively), are defined as follows

$$\begin{cases} \mathbf{P}_{c,max} = \mathbf{P}_{c,min} + \Delta\mathbf{P}_c \\ \mathbf{P}_{d,max} = \mathbf{P}_{d,min} + \Delta\mathbf{P}_d, \end{cases} \tag{17}$$

where, $\Delta\mathbf{P}_c$ and $\Delta\mathbf{P}_d$ are the power margins allocated to different users CUE and DUE (respectively), to ensure that both $\mathbf{P}_{c,max}$ and $\mathbf{P}_{d,max}$ remain greater than $\mathbf{P}_{c,min}$ and $\mathbf{P}_{d,min}$ (respectively). Almost all previous work [6, 8, 9, 11] carried out in this context proposes an amount of powers not to be exceeded for both the CUEs and the DUEs. Since these maximum powers quantities may not verify the criteria already mentioned in the system (1), it is proposed in this chapter to guarantee this condition, by assuming that the power $\mathbf{P_{max}}$ remains always greater than $\mathbf{P_{min}}$. The difference between the two power quantities $\mathbf{P_{max}}$ and $\mathbf{P_{min}}$ is denoted by $\Delta\mathbf{P}_c$ for the CUEs and by $\Delta\mathbf{P}_d$ for the DUEs.

Likewise, we consider the following notation

$$\mathbf{P_{max}} = \left(\underbrace{P_{max}(1), ..P_{max}(K_1)}_{\mathbf{P_{c,max}}}, \underbrace{P_{max}(K_1+1), \dots P_{max}(K)}_{\mathbf{P_{d,max}}} \right). \tag{18}$$

Hence,

$$\mathbf{P}_{max} = (\mathbf{P}_{c,max}, \mathbf{P}_{d,max}). \tag{19}$$

4. On the proposition of a centralized MS and PC approaches

This section investigates centralized MS and PC approaches, which aims to select CUE and DUE from a predetermined list and to minimize the consumed amount of power, in order to satisfy the QoS depicted in Eq. (1). A centralized approach is proposed in this section, which is a generalized version of the algorithm CPCA (denoted GCPCA) to more than one CUE.

The condition assumed during the MS process is $\rho(\mathbf{F}) < 1$. Thus only the users who check this last condition are retained in the final list. Then, the minimum power $\mathbf{P_{min}}$ is allocated to the different types of users (CUE and DUE) based on this selection criterion, in order to optimize the amount of power.

4.1 Proposed generalized centralized power control algorithm (GCPCA)

Unlike the CPCA algorithm which is based on a MS relating to a system containing only one CUE, the GCPCA (see algorithm 1) generalizes this latter for several CUEs, based on the same condition $\mathbf{K_1} \geq 1$. In fact, this assumption is more realistic and illustrates a more real case.

As shown in step 1 from algorithm 1, we first test if the matrix $\mathbf{F}^l$ (relative to the iteration l), verifies the condition $\rho(\mathbf{F^l}) < 1$. If this is true, the Pareto power $\mathbf{P_{min}}$ already defined in the Eq. (11) is assigned to admitted users, as the steps 5 and 6 indicate. Otherwise, we select the $\hat{k}$-th user transmitter (CUE or DUE) who can increase the maximum of interference power compared to other receivers, as shown in step 2. Mathematically, this results in finding user $\hat{k}$, such as: $\hat{k} = argmax_k \|f_k^l\|_2$.

Thus, this user $\hat{k}$ will now be eliminated from the list of users admitted into the system, as is confirmed at step 3.

his most annoying user elimination strategy is repeated until the constraint is verified. Finally, the matrix obtained satisfies the sufficient condition $\rho(\mathbf{F^l}) < 1$, for the existence of the pareto power $\mathbf{P_{min}}$ (see step 5).

All these steps are more detailed in algorithm 1.

Algorithm 1: Proposed GCPCA

---

initialisation: F^l for $l = 0$, for all active CUE and DUE. [Step 1]
Step 1: if $\rho(\mathbf{F}^l) < 1$, go to step 5 and 6. Otherwise, go to Step 2.
Step 2.$\hat{k} = argmax_k \|f_k^l\|_2$
Step 3. remove the $\hat{k}$-th column and row vectors of the matrix $\mathbf{F}^l$.
Step 4. update: $\mathbf{F}^{l+1} = \mathbf{F}^l$, $l = l + 1$. Go to Step 1.
Step 5. evaluate the power $\mathbf{P_{min}}$ using the equation (11).
Step 6. $\mathbf{P} = \mathbf{P_{min}}$

This GCPCA algorithm converges after an iteration number, since the condition $\rho(\mathbf{F}^l) < 1$ must each time be checked by the selected users during each iteration l. In fact, the proposition 2 developed in the previous section provides a convergence certificate of this algorithm.

The maximum power $\mathbf{P}_{max}$ deduced from the previous Eq. (17), will be useful in the next section in order to limit the powers allocated for each type of user.

5. On the proposition of a distributed PC approach based on GT

The power control problem proposed in this paper is considered as a distributed strategies non-cooperative game, where the utility functions as well as the strategies adopted by each user are defined and justified.

5.1 Proposed utility functions

Several utility functions have suggested in [3, 9, 27, 28], using a pricing coefficient to enhance both efficiency and fairness among users. The proposed CUE and DUE utility functions considered in this section are defined as follows [6, 11],

1. **CUE utility function:** The utility function $u_k(\mathbf{P})$ relative to a CUE k is defined as

$$u_k(\mathbf{P}) = -\left(\gamma_k(\mathbf{P}) - \gamma_c^{th}\right)^2, \quad \forall k \in \mathbf{K}_1. \tag{20}$$

2. **DUE utility function:** The utility function $u_k(\mathbf{P})$ relative to a DUE k is defined as

$$u_k(\mathbf{P}) := Rew_k(\mathbf{P}) - Pen_k(\mathbf{P}), \quad \forall k \in \mathbf{K}_2, \tag{21}$$

where

- The reward function $Rew_k(\mathbf{P})$, relative to the kth DUE user, evaluates the payoff of the kth DUE based on both γ_d^{th} and on a nonnegative weighting factor pricing coefficient a_k, as follows

$$Rew_k(\mathbf{P}) = 1 - e^{-a_k\left(\gamma_k(\mathbf{P}) - \gamma_d^{th}\right)}, \quad \forall k \in \mathbf{K}_2, \tag{22}$$

- The penalty function, $Pen_k(\mathbf{P})$, relative to the kth DUE user, is defined as

$$Pen_k(\mathbf{P}) = b_k \frac{C(p_k, \mathbf{P}_{-k})}{I_k(\mathbf{P}_{-k})}, \quad \forall k \in \mathbf{K}_2 \tag{23}$$

where,

$$C(p_k, \mathbf{P}_{-k}) = \sum_{j=1, j\neq k}^{K} p_k g_{k,j} \text{ and } I_k(\mathbf{P}_{-k}) = \sum_{j=1, j\neq k}^{K} p_j g_{k,j} + \sigma^2, \tag{24}$$

and b_k is a constant and nonnegative weighting factor, which reflects the relative impact of the kth DUE user in terms of power. We denote $\mathbf{P}_{-k}$ as the vector of transmit powers of all users other than k, defined as follows

$$\mathbf{P}_{-k} = (p_1, \dots p_{k-1}, p_{k+1}, \dots p_K), \quad \forall k \in \mathbf{K}. \tag{25}$$

From which it follows

$$\mathbf{P} = \left(p_k, \mathbf{P}_{-k}\right). \tag{26}$$

Afterwards, we denote the utility function vector as: $\mathbf{u}(\mathbf{P}) = (u_1(\mathbf{P}), u_2(\mathbf{P}), ..u_K(\mathbf{P}))$, where $u_k(\mathbf{P})$ can be evaluated from (20) or (21), depending on whether the user k is CUE or DUE (respectively).

5.2 Pure strategies game

We denote our game $\mathbf{G} = \{\mathbf{K}, \mathbf{P}, \mathbf{u}(\mathbf{P})\}$ of complete information between $\mathbf{K}$ players. The strategies of such game are considered to be the vector power $\mathbf{P} \in \Omega$, where Ω is given by

$$\Omega = \left\{\mathbf{P} = \left(p_1, p_2, ..p_K\right), p_k \in \Delta_k, \forall k \in \mathbf{K}\right\}, \tag{27}$$

where,

$$\Delta_k = \left[p_{min}(k), p_{max}(k)\right]. \tag{28}$$

The two powers $p_{min}(k)$ and $p_{max}(k)$ are derived from Eqs. (13) and (18) (respectively).

The NE is a strategy profile in which the strategy used by each user is at least as good a reply as any other strategy available to him to the strategies played by the other users. In this sense, to derive the NE of our proposed game, we propose in the following paragraph to study the best response relative to each user k, by improving the utility function of each user (CUE and DUE).

5.3 Nash equilibrium

Definition 2. Best-response: The best-response function $BR_k(\mathbf{P}_{-k})$ of a user k (CUE or DUE) to the profile of strategies $\mathbf{P}_{-k}$, is a set of strategies p_k^* for that user k^* should satisfy the following condition [24]

$$BR_k(\mathbf{P}_{-k}) = \left\{p_k^* \in \Delta_k^*, u_k\left(p_k^*, \mathbf{P}_{-k}\right) \geq u_k\left(p_k, \mathbf{P}_{-k}\right), \forall p_k \in \Delta_k\right\}. \tag{29}$$

Hence, each element of the best-response function $BR_k(\mathbf{P}_{-k})$ is a best-response of the user k, relative to other strategies $\mathbf{P}_{-k}$.

Definition 3. Nash Equilibrium (NE) A pure strategies NE (PSNE) $\mathbf{G} = \{\mathbf{K}, \mathbf{P}, \mathbf{u}(P)\}$, is a set of strategies $\mathbf{P}^* = \left(p_1^*, p_2^*, ...p_K^*\right)$, such that no player can unilaterally enhance its own utility [24], i.e.,

$$u_k\left(p_k^*, \mathbf{P}_{-k}^*\right) \geq u_k\left(p_k, \mathbf{P}_{-k}^*\right), \quad \forall p_k \in \Delta_k. \tag{30}$$

Hence, a PSNE $\mathbf{P}^*$ is a stable outcome of a game $\mathbf{G}$, if no user has any incentive to change its strategy.

5.4 Example: 2-users game

5.4.1 Best-responses expressions of CUE and DUE

We assume that $(K_1 = K_2 = 1)$ and $(a_1 = a_2 = a$ and $b_1 = b_2 = b)$, the expressions of the two Best-responses relatives to CUE and DUE are studied and evaluated in this section. As already explained in the preceding sections, the amount of power for each type of user k should be greater than a minimum power ($p_{c,min} = p_{min}(1)$ for CUE and $p_{d,min} = p_{min}(2)$ for DUE) and also less than a maximum power ($p_{c,max} = p_{max}(1)$ for CUE and $p_{c,max} = p_{max}(2)$ for DUE). Hence, if we denote $\mathbf{P} = (p_1, p_2)$ as the allocated power vector, where p_1 is the power relative to the CUE which should belong to Δ_1 and p_2 is the power relative to the DUE which should belong to Δ_2. We remind that Δ_1 and Δ_2 are already defined in Eq. (28).

In this case, the feasible region of the power is defined as a region where the amount of power $\mathbf{P} = (p_1, p_2) \in \Omega$ should verify the following condition

$$p_{c,\min} \leq p_1 \leq p_{c,\max} \Leftrightarrow p_1 \in \Delta_1 \tag{31}$$

$$p_{d,\min} \leq p_2 \leq p_{d,\max} \Leftrightarrow p_2 \in \Delta_2 \tag{32}$$

Proposition 3. The Best-response relative to the first user (CUE), denoted as $BR_1(p_2)$, is given by

$$BR_1(p_2) = \left\{ p_1 \in \Delta_1, p_1 = \gamma_{th}^c \left(\frac{g_{1,2}}{g_{1,1}} p_2 + \frac{\sigma^2}{g_{1,1}} \right), \quad p_2 \in \Delta_2 \right\}. \tag{33}$$

Proof. Based on the expression of the CUE utility function $u_1(p_1, p_2)$, which is defined in Eq. (20) and where $\gamma_1(\mathbf{P})$ can be derived from the Eq. (1.a), we can easily deduce the following result

$$\frac{\partial u_1(p_1, p_2)}{\partial p_1} = 0 \Rightarrow -2 \frac{g_{1,1}}{g_{1,2} p_2 + \sigma^2} \left(\frac{g_{1,1} p_1}{g_{1,2} p_2 + \sigma^2} - \gamma_{th}^c \right) = 0. \tag{34}$$

So, the expression of $BR_1(p_2)$ found in (33) is derived by deducing the expression of p_1 according to p_2 from the last equation. This completes the proof. □

Proposition 4. The Best-response relative to the second user (DUE), denoted as $BR_2(p_1)$, is given by

$$BR_2(p_1) = \left\{ p_2 \in \Delta_2, p_2 = \frac{g_{2,1} p_1 + \sigma^2}{g_{2,2}} \left[-\frac{1}{a} \log \left(\frac{b g_{1,2}}{a g_{2,2}} \right) + \gamma_{th}^d \right]^+, \quad p_1 \in \Delta_1 \right\}. \tag{35}$$

Proof. Based on the utility function expression $u_2(p_1, p_2)$ defined in Eq. (21) and on the expression of $\gamma_2(\mathbf{P})$ defined in Eq. (1), we can easily get the following expression

$$\frac{\partial u_2(p_1, p_2)}{\partial p_2} = 0 \Rightarrow \frac{a g_{2,2}}{g_{2,1} p_1 + \sigma^2} e^{-a \left(\frac{g_{2,2} p_2}{g_{2,1} p_1 + \sigma^2} - \gamma_{th}^d \right)} - \frac{b g_{1,2}}{g_{2,1} p_1 + \sigma^2} = 0. \tag{36}$$

After simplification, the $BR_2(p_1)$ depicted in Eq. (35) is readily derived by deducing the expression of p_2 according to p_1 from the last equation. □

5.4.2 Simulations and result interpretations

The two best-response $BR_1(p_2)$ and $BR_2(p_1)$ of both CUE and DUE (respectively), which are derived from (33) and (35) (respectively) and the NE are depicted in **Figure 2**. The simulation parameters used in this figure are presented in **Table 1**. As a note from this figure, we can notice that the NE exists and is unique,

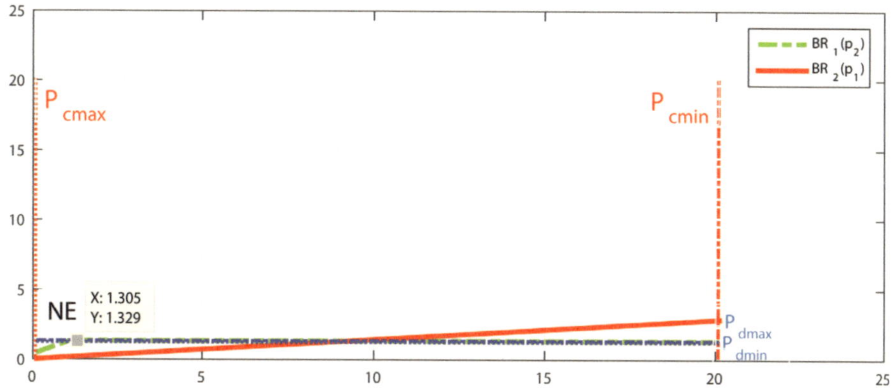

Figure 2.
Best-response $BR_1(p_2)$, $BR_2(p_1)$ and NE.

Parameters	Values
R	700 (m)
D2D link range $d_{k,k}$	50 (m)
D2D link density (λ)	5×10^{-5}
Path loss exponent (α)	4
γ^c_{th}	from -6 to 18 (dB)
γ^d_{th}	from -10 to 14 (dB)
Δp_c	$50mw$
Δp_d	$0.002mw$
σ^2 (for $1MHz$ bandwidth)	$-143.97(dBm)$
	25
b	1
λ	510^{-5}
ε	$1mw$
Iterations number	10^3

Table 1.
Simulation parameters.

because the two curves of $BR_1(p_2)$ and $BR_1(p_2)$ intersect at a single point in the feasible region Ω.

In the next section, we propose to extend this study for a system which contains K_1 CUE and K_2 DUE and furthermore to determine the NE SINR and power closed forms for both CUE and DUE, if it exists.

6. Proposed distributed power control algorithm based on GT

A NE Distributed PC Algorithm (NEDPCA) relative to both CUE and DUE is investigated in this section, in which our proposed game and CUE and DUE utility functions already defined in the previous section are considered. First, to do this, the SINR NE expressions for each user (CUE and DUE) are presented. Afterwards, the amount of power allocated to each user (CUE and DUE) relative to the derived NE are also studied. Thirdly, a power allocation algorithm will be suggested, based on the obtained results to derive the NE power quantities and the power limitation already discussed in Section 3.

6.1 SINR and power NE for CUE and DUE

The authors in [11] have shown that for a CUE and DUE k, the unique SINR NE $\gamma^*_{c,k}(\mathbf{P}^*)$ and $\gamma^*_{d,k}(\mathbf{P}^*)$ (respectively) have the following expressions:

$$\begin{cases} \gamma^*_{c,k}(\mathbf{P}^*) = \gamma^{th}_c, \quad \forall k \in \mathbf{K}_1 \quad (37.a) \\ \gamma^*_{d,k}(\mathbf{P}^*) = \left[\frac{1}{a_k} \log\left(\frac{a_k g_{k,k}}{b_k \sum_{i\neq k} g_{j,k}}\right) + \gamma^{th}_d\right]^+ \quad \forall k \in \mathbf{K}_2, \quad (37.b) \end{cases} \tag{37}$$

where $[f]^+ = max(f, 0)$.

Assumption 1. Based on the Eq. (37.b), we assume the following statement for each DUE k

$$\frac{1}{a_k} \log\left(\frac{a_k g_{k,k}}{b_k \sum_{i\neq k} g_{j,k}}\right) \geq 0 \tag{38}$$

In fact, if: $\frac{1}{a_k} \log\left(\frac{a_k g_{k,k}}{b_k \sum_{i\neq k} g_{j,k}}\right) < 0$, we can have one of the two following cases

$$\begin{cases} \text{case } 1: \text{if } \frac{1}{a_k} \log\left(\frac{a_k g_{k,k}}{b_k \sum_{i\neq k} g_{j,k}}\right) + \gamma^{th}_d \geq 0 \Rightarrow \gamma^*_{d,k}(\mathbf{P}^*) < \gamma^{th}_d. \\ \text{case } 2: \text{if } \frac{1}{a_k} \log\left(\frac{a_k g_{k,k}}{b_k \sum_{i\neq k} g_{j,k}}\right) + \gamma^{th}_d < 0 \Rightarrow \gamma^*_{d,k}(\mathbf{P}^*) = 0. \end{cases} \tag{39}$$

The unique NE power $\mathbf{P}^*$ of both CUE and DUE can be derived as follows

$$\mathbf{P}^* = \left(p^*_1, ..p^*_{K_1}, ...p^*_K\right), \quad \text{where } p^*_k = \left[\frac{I_k(\mathbf{P}^*_{-k})\gamma^*_k}{g_{k,k}}\right]^{p_{max}(k)}_{p_{min}(k)}, \forall k \in \mathbf{K}, \tag{40}$$

where, $\gamma^*_{c,k}(\mathbf{P})$ and $\gamma^*_{d,k}(\mathbf{P})$ are defined in Eq. (37.a) if $k \in \mathbf{K}_1$ and (37.b) if $k \in \mathbf{K}_2$ (respectively).

We propose in the next step a distributed PC algorithm for the mentioned pure strategy game, which is based on the allocated power $\mathbf{P}^*$ previously defined in Eq. (40).

6.2 Proposed distributed power control algorithm based on GT

The algorithm depicted in 2 outlines the different steps of the proposed algorithm NEDPCA offered to each CUE and DUE, which is based on the previous pure strategy game. In fact, this algorithm NEDPCA offers a NE power for the two kinds of users, based on both the previous constraints (1) and on the power expression depicted in Eq. (40). First, the MS process is derived from the GCPCA, in order to guarantee the constraints imposed by the CUEs and the DUEs in terms of SINR thresholds (see Eq. (1)). This is shown in step 1 of the Algorithm 2. Second, the SINR NE $\gamma^*_{c,k}(\mathbf{P})$ and $\gamma^*_{d,k}(\mathbf{P})$ relative each user k (CUE and DUE) are obtained, which are evaluated in Eqs. (37.a) and (37.b) (respectively) (as it is shown from steps 2 and 3). Third, the PC process is investigated based on the iterative approach which is executed by using Eq. (40), as shown in steps 4 and 5. Step 6 allows to finalize the power distribution step, with an error of nearly δ for each user k. If $|p_k^{*(t+1} - p_k^{*t}| < \delta\ \forall$ user k, then the amount of power allocated to each user k, is given as: $\mathbf{P}_k^{*t} = \mathbf{P}_k^*$, otherwise we repeat this process and go to step 2.

Algorithm 2: Proposed NEDPCA

---

initialisation: $t = 0, \mathbf{F}, \mathbf{b}, \delta > 0$ and $p_k = p_k^0, \forall k$ user.
Step 1: Evaluate from the Algorithm 1 GCPCA:
  1) the CUE and DUE set of users: $\mathbf{K_1}$, $\mathbf{K_2}$
  2) $\mathbf{P}_{min}$ and $\mathbf{P}_{max}$ using equations (17) and (19).
  2. for each CUE k, evaluate $\gamma^*_{c,k}(P)$ using equation (37).a.
Step 3. for each DUE k, evaluate $\gamma^*_{d,k}(\mathbf{P})$ using equation (37).b.
Step 4. Derive for each CUE and DUE k, the amount of power p_k^{*t} using equation (40), where $p_{min}(k)$ and $p_{max}(k)$ are derived from the 1st step. Evaluate $\mathbf{P}^{*t} = \left(p_1^{*t}, ..p_K^{*t}\right)$
Step 5: update $\mathbf{P}^{*(t+1)} = \mathbf{F}\mathbf{P}^{*t} + \mathbf{b}$.
Step 6. if $|p_k^{*\left(t+1-p_k^{*t}\right| < \delta, \forall k}$, derive the solution $\mathbf{P}_k^{*t} = \mathbf{P}_k^*$, otherwise $t = t + 1$ and go to the 2st step.

All the NEDPCA steps relative to the distributed MS and PC for both CUE and DUE are detailed in Algorithm 2. Indeed, the first step of NECPCA makes it possible to realize the MS approach and all the other steps allow to deduce the PC approach.

Like the GCPCA algorithm, the NEDPCA algorithm converges after an iteration number, since it is based on the same condition $\rho(\mathbf{F}^l) < 1$ which must be checked during each iteration l by all the selected users. In fact, by applying the step 1 of the algorithm GCPCA (see Algorithm 2), the last condition should be guaranteed. It is also due to the proposition 2, that the convergence of NEDPCA is proved.

7. Analysis of simulations

In order to evaluate the performance of the algorithms already mentioned and proposed in the following sections, we consider in this section to study the simulations of these algorithms: GCPCA and NEDPCA. A Monte Carlo simulation is applied according to the Table 1, already given in the previous section.

The CUE and DUE Total powers are evaluated in **Figures 3** and **4** versus γ_c^{th} and γ_d^{th} (respectively).

We remain that the CPCA algorithm considers only one CUE and possibly several DUE. The GCPCA allocates to the different users the minimum power derived from Eq. (11), while respecting the condition $\rho(\mathbf{F}) < 1$. Thereby, any CUE and/or DUE that does not verify this condition will be eliminated from the user list.

First, we can notice from **Figure 3** that all the curves relative to GCPCA and NEDPCA algorithms are decreasing when γ_c^{th} increases. This is because when the threshold γ_c^{th} increases, it becomes difficult to find CUEs that check the constraints depicted in (1). On the other side from **Figure 4**, only the DUE performances are improved when GCPCA is considered and γ_d^{th} is improved. This is due to the fact that, the NE allows to maximize the utility function relative to each user k.

As it is shown from these two figures, the DUE coverage probability is significantly improved compared to that of DUE, because the DUEs benefit much more from TG compared to CUE, by using the NEPCA.

Second, the proposed centralized approach GCPCA which is a generalization of the CPCA approach offers less total power compared to the NEPCA, for both types of users CUE and DUE. This is due to the fact that when we want to reach a NE, by increasing the utility functions of each type of user, we should increase the total power consumed. Moreover, the difference between the two types of curves represents the power gain that must be added in order to reach this NE. From **Figures 3** and **4**, it is clear that for DUEs, this difference in terms of power is smaller

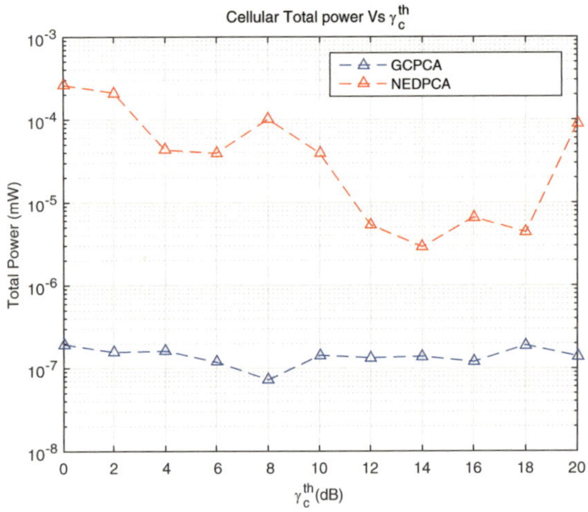

Figure 3.
CUE Total power vs. γ_c^{th}.

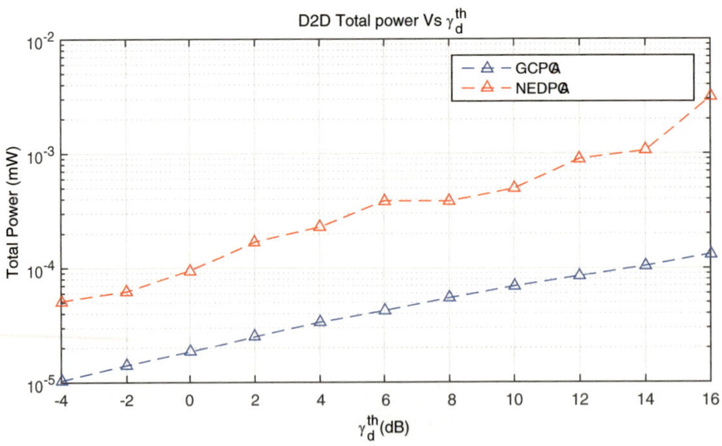

Figure 4.
DUE Total power vs. γ_d^{th}.

compared to that of the CUEs. This is explained by the fact that the DUEs require less power to transmit. Third, by limiting the power consumed in $[\mathbf{P}_{min}, \mathbf{P}_{max}]$ when GCPCA and NEDPCA are used, more flexibility and possibility to integrate the two types of users are offered. The greater the amount of maximum power $\mathbf{P}_{max}$, the higher the probability of coverage using the proposed GCPCA algorithm compared to CPCA one. It is for these reasons that it would be judicious to choose adequate margins of power $\Delta\mathbf{P}_c$ and $\Delta\mathbf{P}_d$, relative to the types of users (CUE and DUE).

8. Conclusions

This chapter allows to invoke the problem of selection mode and power control for a D2D underlaid cellular networks in 5G. The basic idea of this chapter is to generalize the classic allocation algorithms by applying Game Theory, for many CUEs and DUEs in system.

First, we assume that the amount of power allocated to each kind of user should be between two amounts of power: a minimum power defined as a Pareto solution and a maximum power. Thus, a mathematical demonstration was provided in this chapter, in order to prove the expressions of these two powers, based on constraints imposed by the users in terms of SINR thresholds to be respected.

Second, our proposed system is modeled as a non-cooperative pure game between the different types of users, where the utility functions should be maximized. From the built-in utility functions, NE SINR and PC solution existence and uniqueness are investigated and studied.

Third, simulations have been established in this context, in order to assess the performance of the algorithms thus proposed in terms of total powers relative to both users CUE and DUE. Through these simulations which compare these results without and with GT, we noticed that by applying the TG, the total power consumed increases in order to reach the NE for the two types of users: CUE and DUE. This is due to the fact that to increase the utility functions relating to the two types of users, a power margin must be added. However, this difference in terms of power becomes less important for the DUEs, since they require less power relative to the DUEs.

Author details

Sameh Najeh[1*] and Ammar Bouallegue[2]

1 Higher School of Communications of Tunis, COSIM Research Laboratory, University of Carthage, Tunis, Tunisia

2 National Engineering School of Tunis, SysCom Research Laboratory, University of Tunis - El Manar, Tunis, Tunisia

*Address all correspondence to: sameh.najeh@supcom.tn

References

[1] Simsek M, Aijaz A, Dohler M, Sachs J, Fettweis G. 5G-enabled tactile internet. IEEE Journal on Selected Areas in Communications. 2016;**34**(3): 460-473

[2] 3GPP. Evolved Universal Terrestrial Radio Access (E-UTRA) and Evolved Universal Terrestrial Radio Access Network (E-UTRAN); Overall description. TS 36.300 (Rel. 15); 2018

[3] Gao C, Li Y, Zhao Y, Chen S. A two-level game theory approach for joint relay selection and resource allocation in network coding assisted D2D communications. IEEE Transactions on Mobile Computing. 2017;**16**(10): 2697-2711

[4] Melki L, Najeh S, Besbes H. Interference management scheme for network-assisted multi-hop D2D communications. In: IEEE International Symposium on Personal, Indoor and Mobile Radio Communications (PIMRC); Spain; 2016. pp. 1-5. DOI: 10.1109/PIMRC.2016.7794834

[5] Melki L, Najeh S, Besbes H. Radio resource management and outage analysis for network-assisted multi-hop D2D communications. Digital Communications and Networks. 2016; **2**(4):225-232

[6] Lee N, Lin X, Andrews JG, Heath RW Jr. Power control for D2D underlaid cellular networks: Modeling, algorithms and analysis. IEEE Journal on Selected Areas in Communications. 2015;**33**(1): 1-13

[7] Katsinis G, Tsiropoulou E, Papavassiliou S. Multicell interference management in device to device underlay cellular networks. Future Internet. 2017;**9**(3):1-20

[8] Ali KS, ElSawy H, Alouini MS. Selection and power control for uplink D2D communication in cellular networks. In: International Conference on Communication Workshop (ICCW). IEEE; 2015. pp. 620-626

[9] Ye Q, Al-Shalash M, Caramanis C, Andrews JG. Resource optimization in device-to-device cellular systems using time-frequency hopping. IEEE Transactions on Wireless Communications. 2014;**13**(10): 5467-5480

[10] Koskie S, Gajic Z. Nash game algorithm for SIR-based power control in 3G wireless CDMA networks. IEEE Transactions on Networkings. 2005; **13**(5):1017-1026

[11] Najeh S, Bouallegue A. Game theory for SINR-based power control in device-to-device communications. Physical Communication. 2019;**34**:135-143

[12] Najeh S, Bouallegue A. Distributed vs centralized game theory-based mode selection and power control for D2D communications. Physical Communication. 2019;**38**, **1**:-10

[13] Zhang K, Geng X. Power control in D2D network based on game theory. In: International Conference on Intelligence Science. Shanghai, China; 2017. pp. 104-112

[14] Lashgari M, Maham B, Kebriaei H. Energy-efficient self-backhauling in heterogeneous wireless networks: A game-theoretic approach. Physical Communication. 2018;**29**:296-306

[15] Baniasadi M, Maham B, Kebriaei H. Power control for D2D underlay cellular communication: Game theory approach. In: International Symposium on Telecommunications (IST); 2016. DOI: 10.1109/ISTEL.2016.7881832

[16] Kebriaei H, Maham B, Niyato N. Double sided bandwidth-auction game

for cognitive device-to-device communication in cellular networks. In: IEEE Transactions on Vehicular Technology; 2015. DOI: 10.1109/TVT.2015.2485304

[17] Banagar M, Maham B, Popovski P, Pantisano F. Power distribution of device-to-device Communications in Underlaid Cellular Networks. IEEE Wireless Communications Letters. 2016; **5**(2):204-207

[18] Doppler K, Yu C, Ribeiro C, Janis P. Mode selection for device-to-device communication underlaying an lte-advanced network. In: IEEE Wireless Communications and Networking Conference (WCNC); 2010. pp. 1-6

[19] Abdallah A, Mansour MM, Cheha A. A distance-based power control scheme for D2D communications using stochastic geometry. In: VTC-Fall; 2017. pp. 1-6

[20] Rashed SK, Shahbazian R, Ghorash SA. Learning-based resource allocation in D2D communications with QoS and fairness considerations. Transactions on Emerging Telecommunications Technologies. 2017;**29**(1):1-6

[21] Wang R, Zhang J, Letaiefa KB. Incentive mechanism design for cache-assisted D2D communications: A mobility-aware approach. In: International Workshop on Signal Processing Advances in Wireless Communications (SPAWC). Japan; 2017. pp. 1-5. DOI: 10.1109/SPAWC.2017.8227678

[22] Trigui I, Affes S. Unified analysis and optimization of D2D communications in cellular networks over fading channels. IEEE Transactions on Communications. 2018;**67**:724-736

[23] Neumann JV, Morgenstern O. Theory of Games and Economic Behavior. United States: Princeton University Press; 1944. ISBN: 978-0691130613

[24] Han Z, Niyato D, Saad W, Baar T, Hjrungnes A. Game Theory in Wireless and Communication Networks: Theory, Models, and Applications. Cambridge University Press; 2011. DOI: 10.1017/CBO9780511895043

[25] Ibrahim R, Assaad M, Sayrac B, Gati A. When distributed outperforms centralized scheduling in D2D-enabled cellular networks. In: 21st ACM International Conference; 2018. DOI: 10.1145/3242102.3242115

[26] Foschini G, Miljanic Z. A simple distributed autonomous power control algorithm and its convergence. IEEE Transactions on Vehicular Technology. 1993;**42**(4):641-646

[27] Zappone A, Sanguinetti L, Bacci G, Jorswiec E, Debbah M. Energy-efficient power control: A look at 5G wireless technologies. IEEE Transactions on Signal Processing. 2016;**64**:1668-1683

[28] Zhu Q, Pavel L. Stackelberg game approach to constrained OSNR Nash Game in WDM optical networks. In: American Control Conference; 2008. pp. 1967-1972

Chapter 13

Perspective about Medicine Problems via Mathematical Game Theory: An Overview

Agostino Bruzzone and Lucia Pusillo

Abstract

This chapter provides an overview of Game Theory with applications to medicine problems, including evolution of tumor cells and their competition, applications to neocortical epilepsy surgery and schizophrenic brain. Recent studies related to microarray games for cancer problems will be considered. These models may be used for applications to neurological and allergic diseases. At the end, the model of kidney exchange via the Matching Theory proposed by Alvin Roth, Nobel prize 2012, will be discussed.

Keywords: game theory, Nash equilibria, evolutionary stable strategy, tumor cells, thoracic surgery, neocortical epilepsy, kidney donation

1. Introduction

Mathematical Game Theory is a branch of Mathematics which analyzes strategic behaviors of decision makers which interact each other: the players. They have cooperation or competition each other. Game Theory started in 1944 with J.von Neumann and O.Morgenstein with the book which pick up their researches "Theory of Games and Economic Behavior". They started studying only zero sum games (intuitively if a player wins the other loses) and thanks to J.Nash, another giant in Games Theory, the general games (not necessary zero sum) were introduced and studied.

There are two great chapters in this science: cooperative games and noncooperative games. There is a "bridge" between cooperative and noncooperative games and it consists of the "repeated games". In fact if a game is repeated with infinite horizon the solution is in the cooperative zone (Folk Theorem) (see [1]).

In this contribution we consider medicine problems modeled by mathematical games. We present some models as evolution of cancer cells and their competition, applications to neocortical epilepsy surgery and schizophrenic brain. Some studies which we will see in details in the next sections with needed references, keep into account the microarray games which were studied to consider the cancer problem but applied also to neurological and allergic diseases. The model of kidney exchange is presented too, this is studied via an algorithm proposed by A.Roth, Nobel prize 2012. Also some management problems of modern health care are considered. Even if Game Theory is a young science, in many situations has revealed to be a fundamental tool as for the genetic analysis or to study images to determine cortical surface displacement during a surgery based on intraoperative stereo image information.

Furthermore studying the strategic interactions, this science permits us to investigate the relations between sick cells and healthy ones and among various genes.

We wish note that the examples are made with two or three players, for easy notations but they can generalize to a large number of players.

An interesting open problem is to apply the simulation techniques used in [2–5] and adapt them for simulations in medicine problems. Furthermore we can use the strategic method to evaluate behaviors for example in a military problem, and apply this model to evaluate the dangerous cells in a suspected disease.

This chapter is organized as follows: in Section 2 we analyze the Doctor Dilemma which is a generalization of the Prisoner Dilemma, in Section 3 we consider Evolutionary Game Theory, branch of Game Theory that can be successfully applied to Medicine models. In Section 4 we consider the evolution of cancer cells presenting many game models. In Section 5 we study the applications to brain models and in Section 6 we analyze the interesting problems of kidney exchange via the Matching Theory proposed by Alvin Roth, who won the Nobel Prize in 2012. Conclude the paper a rich but not exhaustive bibliography which invites to further reading of game theoretical texts.

2. Prisoner dilemma and doctor dilemma

We think that Game Theory offers to Medical Science a new interesting method to analyze and model some problems.

There are many papers about these two sciences studied together but for the complexity of the problem we can say that we are at the beginning of a new challenge team work.

Relieving pain and suffering is one of the most important doctors roles, so these have much gratification when they achieve this. Society is expecting that doctors successfully manage pain in fact this is one criterion that the Joint Commission certifies as health care organization in the United States. Given the relevance of trust and cooperation among patients and doctors for health care we can apply Game Theory to study the interaction in health settings and to understand the strategies which must be adopted to have the best payoffs for doctor and his/her patients [6].

In addition physician's compensation may depend on patients' satisfaction. But some problems might be: for example the patient could have no pain but he/she wants drugs for other motivations. How should the doctor do? This problem may be modeling as the Prisoner Dilemma where the players are the patient (player I) and the doctor (player II) and the strategies are respectively $X = \{real\ pain,\ fake\ pain\}$ and $Y = \{prescribe\ drugs,\ do\ not\ precibe\ drugs\}$.

This is the Doctor dilemma game where the best outcomes for both the players are not the equilibrium of the game.

Doctor patient	Prescribe drugs	Do not prescribe drugs
Real pain	Patient satisfied, High satisfaction score for the doctor, Professionally rewarding	The patient dissatisfied, Low satisfaction for the doctor
Fake pain	Patient satisfied, High satisfaction for the doctor, Even if it is professionally less rewarding	The patient dissatisfied, Low satisfaction for the doctor, Even if it is professionally most rewarding

This game is similar, even if it is not symmetric, to the Prisoner Dilemma below.

$I\backslash II$	L	R
T	1, 1	5, 0
B	0, 5	3, 3

In this game $X = \{T, B\}$ and $Y = \{L, R\}$ are the strategy spaces of player I and player II respectively. T is for Top, B is for Bottom, L is for Left and R is for Right, but these are only names.

The *NE* is (T, L) in fact.

$T = R_I(L)$ the best reply of player I to the strategy L of player II and.

$L = R_{II}(T)$ the best reply of player II to the strategy T of player I.

It is a nonefficient solution but in dominant strategies.

The situation is very complex: the physician cannot be sure if the patient tells true or not, the patient is in a position of vulnerability, the physician cannot guarantee the successful of his/her prescription (if any). The physician can fall prescribing drugs if they are not necessary or also necessary if these have collateral dangerous effects for the health patient. This wishes the drugs and if the doctor does not prescribe them, his/her trust falls down. There are a lot of problems in these decisions, some are discussed in [7].

Another interesting medicine problem is studied in [8], where the "consultation games" are introduced. The author considers important the cooperation between patients and doctors by building trust, obtaining information and solving problems. The physician payoffs depends on proposing care methods and communicating with patients. The payoff of patients depends sometimes from a fast provision of medical interview and sometimes from relationship with the doctor (in special way in case of chronic illness). Other patients wish a screening investigation but they do not know the possible harms involved. Also in these cases the games are useful to understand the best strategies.

In [9], the authors propose noncooperative games for surgery problems and cooperative games for the operating room settings to create a better synergy and improve the hospital efficiency and patients safety.

A reiterated process leads us to make a model via repeated games (see [10, 11] or through the combination of cooperative and not cooperative games. In real situations not all the players wish to cooperate: some of them cooperate but some else do not. In this case we speak about partially cooperative games, studied in [12].

3. Evolutionary game theory

A very interesting chapter is Evolutionary Game Theory proposed by Maynard Smith, a theoretical evolutionary biologist, geneticist and an aeronautical engineer to study mathematical models in biology and its strategic aspects.

Many socioeconomic and biological processes can be modeled with interacting individuals where players wish to maximize their own payoffs and in particular animals and genes maximize their individual fitness, [13, 14].

In evolutionistic selection, individuals wish to maximize the expected value of a measure of surviving otherwise they are substituted, to this goal they decide their strategies, not consciously but following evolution rules. The concept of human rationality is substituted by evaluative fitness, and Maynard Smith had many doubts on the first but no on the second.

We can apply this theory to understand the cancer cells propagation. In an evolutionary context we consider animals or bacteria players as maximizing their fitness so we consider steady state population dynamics. Intuitively each evolutionary player uses a strategy to maximize its payoff following the evolutionary theory and via a dynamical convergence to a stable outcome or the so called evolutionary stable strategy (*ESS* for short). An equilibrium deriving from an *ESS* is a refinement of a Nash equilibrium (*NE*) and it is a stable solution under small perturbations.

We will write the following example as a generalization of the well-known Hawk/Dove game.

In a human body there are 2 groups of cells: healthy cells and cancer cells, they want to "conquer" the human body and so they must fight. They can have a quiet behavior (D) or an aggressive behavior (H) and this is decisive to have the spreading of the disease or not. We call c the cost of the fight and it depends from the resources body and the medicines given to the patient when the disease was discovered.

This game is written in the following matrix:

I\II	*H*	*D*
H	$1/2 - c,\ 1/2 - c$	1, 0
D	0, 1	1/2, 1/2

Player I and player II are animals in the original game (here we can think to two patients).

If both the animals have a pacific behavior, they divide the "prey", so their payoff is $1/2$ (the individual is not completely ill).

If the behavior of a group of cells is H and the other is$D$, the aggressive ones invade the human body and the other does not.

So the patient has the disease if the aggressiveness is from tumor cells instead he/she is recruited if it derives from healthy cells. If both the cells are aggressive, they obtain $1/2 - c$: they divide the human body but they have a detriment of strength (equal to cost c for the fight).

Studying the *ESS* and the subsequent equilibria we can prove:

if $c \leq 1/2$ there is only one equilibrium from *ESS*: (H, H).

if $c > 1/2$ there are no evolutionary equilibria from *ESS*.

This result is intuitive in fact if the fight cost is high (intuitively greater than 1/2) it is better do not fight and leave the prey (in our case the human body in its state).

This new theory may be applied to spatial stress variations, such as the case in cancer dynamics. In [10] the authors studied a game theoretical model in the dynamics of cooperators and cheater cells under metabolic stress hypotheses and spatial heterogeneity. Via Game Theory they tried to understand the dynamic of cancer tumor evolution under stress. They give a simulation of the development of cancer cells under the hypotheses of exchange of genetic material between the individuals (this process is called "horizontal gene transfer"). The authors suppose individuals can change their strategies from being cheaters to cooperators. A strategy can evolve by genetic evolution as a reply to the stress of the local environment. A combined dialog between these models and lab experiments can enrich our knowledge about tumor cells resistance.

To read further research about Evolutionary Game Theory and cancer, see [15–18].

Sometimes cancer cells in the primary tumor may stop growing (for example for restricted space or few oxygen) and some malignant cells may break away from the primary tumor. There are a lot of open questions: Which kind of tumor cells and how much go far from the primary tumor? Which of them migrate into the vessels? Which kind of circulating tumor cells and how many go in a second tumor? A games model which keeps into account the interactions among them and the environment, could be very useful. See [19].

4. Evolution of tumor cells

An unusual hypothesis about the cooperation of tumor cells may be found in [20]. The authors think tumor cells as players in a mathematical game, their interactions (intuitively their strategies even if not consciously decided), permit them to arrive to evolutionist fitness. Distinct tumor cells cooperate to overcome some host defenses by exchanging different products. Two nearby subclones can protect each other spending the process of tumorigenesis, thanks to malignant cells containing all nourishment for cancer growing.

Another model studying the cancer cells was introduced in [21] and generalized to multicriteria games (that is games with vector payoffs) in [22].

The authors consider, via mathematical game theory, the genic expression to investigate serious diseases as cancer. Their goal is to propose a method for evaluating the relevance of the genes as disease markers. The common application in Medicine is "to teach" a "classificator" to distinguish between healthy and sick subjects on the basis of samples given by doctors.

A method to make a feature selection is to use Cooperative Game Theory with transferable utility (*TU*-games in literature, see [10, 11].

Intuitively, each gene is considered in a coalition to which a value is assigned and it shows how much these genes expressions suggest to distinguish between healthy and sick subjects. In the cited paper, the authors applied mathematical Game Theory to analyze the results obtained with microarray techniques which allow to make a photo of thousand of genes expressions through a unique experiment. The starting point is studying the genetic expression in a cell sample and verifying some particular biological conditions (for example the cells of a subject affected by a tumoral disease). Studying the "microarray games" we can evaluate the relevance of genes to regulate or to provoke the onset of a pathology taking into account the genes interactions each other.

In this context the Shapley and the Banzhaf values (see [10, 11] are studied as a measure about the "importance" of a gene ("relevance index") in the diagnosis.

The authors in [22] study the vector Shapley value for microarray multiobjective games basing on the idea of "partnership of genes" (as in [21]). Intuitively this is a genes' group with correlated characterizations and which is very useful to study if the disease is developing.

The experimental results have shown that the Shapley value is a valid tool to evaluate the expressions of genes and to predict a tumor disease.

The advantage of considering a coalitional game is the possibility to compute a numerical index, the so called *relevance index* which intuitively represents the relevance of each gene taking into account the relevance of the others when, for example, a tumor disease is growing.

In general is important to consider multiobjective games (or vectorial games) instead of scalar ones, because the players have not one but more objective "to maximize" and often these goals are not comparable. In a medicine problems there are many parameters "to optimize" so it is important to keep into account them together, the solution will be more precise and it will allow us to better understand the presence of a disease.

About multiobjective games and their solutions you can see: [12, 23–25].

In the following example we consider two real expression matrices where the first M^{S_R} contains the samples of patients without considered disease, the second M^{S_D} contains samples of patients who have some disease signals, so we must investigate about two medical parameters in the set of genes $\{1, 2, 3\}$ which are considered the players in our model. The study is made w.r.t. the samples $\{a, b, c, d\}$ in the case of patients without disease and w.r.t. the samples $\{a, b, c\}$ in the case of patients with disease signals. After a comparison between the two matrices M^{S_R} and M^{S_D} written below, we arrive to write the microarray cooperative game for our model.

The two objectives may be for example the values of protein index and glycemic index. If we will study more objectives (suggested by the doctors), our model will be more precise.

M^{S_R}

	Sample a	**Sample b**	**Sample c**	**Sample d**
Gene 1	$\begin{pmatrix} 7.77 \\ 0.5 \end{pmatrix}$	$\begin{pmatrix} 8.95 \\ 0.2 \end{pmatrix}$	$\begin{pmatrix} 6.48 \\ 0.3 \end{pmatrix}$	$\begin{pmatrix} 1.94 \\ 0.6 \end{pmatrix}$
Gene 2	$\begin{pmatrix} 20,40 \\ 12 \end{pmatrix}$	$\begin{pmatrix} 14,75 \\ 10 \end{pmatrix}$	$\begin{pmatrix} 34.88 \\ 4 \end{pmatrix}$	$\begin{pmatrix} 20.35 \\ 5 \end{pmatrix}$
Gene 3	$\begin{pmatrix} 0.49 \\ 8 \end{pmatrix}$	$\begin{pmatrix} 5.79 \\ 13 \end{pmatrix}$	$\begin{pmatrix} 1.00 \\ 20 \end{pmatrix}$	$\begin{pmatrix} 16.47 \\ 9 \end{pmatrix}$

M^{S_D}

	Sample a	**Sample b**	**Sample c**
Gene 1	$\begin{pmatrix} 3.26 \\ 0.9 \end{pmatrix}$	$\begin{pmatrix} 1.63 \\ 0.4 \end{pmatrix}$	$\begin{pmatrix} 1.58 \\ 0.7 \end{pmatrix}$
Gene 2	$\begin{pmatrix} 89.52 \\ 4.6 \end{pmatrix}$	$\begin{pmatrix} 17.35 \\ 11 \end{pmatrix}$	$\begin{pmatrix} 15.76 \\ 18 \end{pmatrix}$
Gene 3	$\begin{pmatrix} 4.66 \\ 11 \end{pmatrix}$	$\begin{pmatrix} 0.3 \\ 3321 \end{pmatrix}$	$\begin{pmatrix} 19.44 \\ 12 \end{pmatrix}$

From these two tables we can build a cooperative microarray game (with two objectives) which is:

S	∅	{1}	{2}	{3}	{1,2}	{1,3}	{2,3}	{1,2,3}
$v(S)$	0	0	$\frac{1}{3}$	0	$\frac{1}{3}$	$\frac{2}{3}$	$\frac{1}{3}$	1
	0	$\frac{1}{3}$	0	$\frac{1}{3}$	$\frac{2}{3}$	$\frac{2}{3}$	$\frac{1}{3}$	1

So the physician problem studied through microarrays analysis may be translated into a cooperative game and studied with the usual tools of the theory (for more and not easy details see [22]). Here the authors consider the vector Shapley value and the Banzhaf one as the genes relevance indices or disease markers but a research is in progress about other solutions perhaps more suitable and a comparison among the results.

5. Game theory and brain

In some papers of Game Theory diseases connected with our brain are considered, also microarray games may be applied to study neurological diseases, see [26].

In the paper [27], the author consider Game Theory in relation to schizophrenic brain. The idea was emerging from the disease of John Nash. He won the Nobel prize in 1994 for his fundamental works about the concept of equilibrium (the so called Nash equilibrium), the existence of at least one equilibrium in mixed strategies in finite games and the existence of an equilibrium in bargaining games [10, 11].

The disease of John Nash was discovered when he was 29 years old and a brilliant mathematics researcher. He generalized the games from a zero sum (introduced by von Neumann and Morgestein) to general games. This is very important, because in the real life only a few games are zero sum that is a player wins and the other loses, but everyone wins something.

The author applies the theory of strategic interaction to understand the behavior of schizophrenic brain, suggesting us the study between the limbic system and the cerebral hemispheres and between the two cerebral hemispheres. We have an equilibrium when the brain is optimally working, that is when an hemisphere is not prevalent on the other but they are in equilibrium. In our model of a noncooperative game the two cerebral hemispheres are seen as two players playing against each other. This is possible because, as schizophrenic brain researcher teaches us, the two hemispheres are isolated one from other or because there is too much connection between them. In any case the effect is disastrous. Also the relations between the lower brain functions may be seen in the same way. Following the research about schizophrenic disease one hemisphere does not interfere with the optimal functioning of the other.

In the Prisoner Dilemma the two players decide contemporary and independently their strategies that is no one knows what the other does, so it may be a good model for a schizophrenic brain and to understand better this terrible disease.

Games are used also for applications to neocortical epilepsy surgery, see [28] where this topic is applied to cortical surface tracking during the neurosurgery to have information about the brain surface deformation and to have a good image. This method has a high percent of success. Surprisingly this young science is applied also to the image analysis, [29, 30] and to estimate the brain deformation.

Deformations are studied as strategies in a noncooperative game.

The goal is the research of a *NE* which is a strategy profile where there is no incentive for players to deviate unilaterally from their strategy.

In the surgery game we consider two hypothetical players whose strategies are respectively the dense displacement field and the camera calibration parameters which are used to transform points from the 3D intraoperative field into stereo image space. The utility functions are the opposite of the cost functions corresponding to the values for the operations.

All this can be formulated as a noncooperative game in fact changing the values of the camera calibration parameters, also the search of displacement field may vary. The authors use a game theoretic algorithm based on a Bayesian approach to have information about the cortical surface deformation.

A paper about neuroscience is [31] where combining the modern neuroscience methods and mathematical games, a neuroeconomics approach, studying the knowledge of brain mechanism, helps us in keeping social decisions.

6. The matching theory and kidney exchanges

Sometimes, many scientific discoveries are not easy to support economically or awake bioethics problems so these may not be easily applied to disease patients. This problem happens, for example, in the kidney exchanges. In this exchange the successful has arrived to $95 - 97\%$ and the sensibility of giving a transplant organ after death has grown. In despite to these great successful there are many patients waiting an organ transplant because there are not a sufficient number of organs to give to needing patients.

In the sub Saharan Africa, each years a lot of people (more than 5 million) die because they have no admission neither to hemodialysis nor kidney transplant. There is another great problem, which is incompatibility between donors and recipients. In the most number of countries the unique admissible solution by law is the existence a family degree between donor and recipient. Sometimes donor and recipient are not compatible because their blood group or other characteristics of their immune system.

In 2012 Alvin Roth, professor at the Stanford University, won the Nobel prize for his Matching Theory which started from a kidney exchange problem [32, 33]. To explain intuitively this theory we make a simple example: let us suppose that a father needs a kidney transplant, because in a few week he risks to die, and her daughter wishes to give it to him but she is not compatible. In the world there is certainly another pair in the same situation. So the Matching Theory suggests to create a special chain among people who wants give a kidney to a dear person but because incompatibility this is not possible. Starting from Matching Theory a great project presented by Alvin Roth and other scientists, the Global Kidney Exchange Program, this is an international system with donors and recipients with compatible organs. An algorithm proposed by Roth, tests the pair in the program and finds the best match between waiting patients and donors so the patients may find a compatible organ in time for their survival. This program is very important not only because it gives a good life to a person but also for the economic tolerability of the sanitarian system. One objective is to help people needing kidney exchange in the poor countries (where there is no hemodialysis). If a kidney for an Italian patient comes from an African patient, this will have a new kidney and he will live and our Italian patient will have a better life with less cost for the National Sanitarian Service. Keeping into account the compatibility between pair of donors and receivers, the problem may be in great part resolved. Making the mathematical model via a game, we can prove that the *NE* is efficient because all players obtain the best.

As you can imagine there are some ethical problems to consider in fact in an international chain some unscrupulously people could enter and the problem of organ traffic must be supervised. The Global Kidney Exchange has had the endorsement of the American Society of Transplant Surgeons and the World Health Organization has promoted this challenge idea.

(See https://www.profignaziomarino.com/mc/468/trapianti-d-organo-la-proposta-rivoluzionaria-di-un-nobel). About kidney exchange and Game Theory see also [34] and references in it.

7. Conclusion

The overlapping results obtained on mathematical games and those in medical literature may not be casual, medicine models studying the disease onset could make use of methods of this young science. Once again we understand that scientific research need of a studious team in many science branches: mathematicians, engineers, medicine, biologists, economists and so on.

Mathematical games contribute to suggest us the ways to verify new ideas and they permit us to prove with precise calculations the logic of our reasoning because the better understanding of these diseases is an important step.

We remark that for all notions which we have not defined in the text, we refer to already cited [10, 11, 13, 35] the book of Roger Myerson, Nobel Prize 2007 for the theory of Mechanism Design.

Author details

Agostino Bruzzone[1] and Lucia Pusillo[2]*

1 DIME - Department of Mechanical, Energy, Management and Transportation Engineering, University of Genova, Genova, Italy

2 DIMA - Department of Mathematics, University of Genova, Genova, Italy

*Address all correspondence to: pusillo@dima.unige.it; luciapusillo@yahoo.it

References

[1] Binmore K., Fun and Games. A text on Game Theory. D.C. Heath and Company. Lexington, Massachusetts, Toronto. 1992

[2] Bruzzone A, Fadda P, Fancello G et al. Logistics node simulator as an enabler for supply chain development: innovative portainer simulator as the assessment tool for human factors in port cranes. Simulation. 2011;87(10): 857–874.

[3] Bruzzone A. Intelligent agent-based simulation for supporting operational planning in country reconstruction. International Journal of Simulation and Process Modelling. 2013;8(2–3):145–159.

[4] Bruzzone A, Massei M, Tremori A et al. Simulation as enabling technologies for agile thinking: training and education aids for decision makers. International Journal of Simulation and Process Modelling. 2014;9(1–2):113–127.

[5] Bruzzone A, Massei M. Simulation-Based Military Training. In: Mittal S, Durak U, Ören T. editors. Guide to Simulation-Based Disciplines. Springer; 2017. p. 315–361.

[6] Tarrant C., Dixon Woods M., Colman Andrew M.,Stokes T., Continuity of trust in Primary Care: A Qualitative Study Informed by Game Theory. Ann. of Family Medicine, 2010;8:5

[7] Djulbegovic B, Hozo I. When is it rational to partecipate in a clinical trial? A game theory approach incorporating trust, regret and guilt. BMC Med Res Methodol 2012;12:85

[8] Elwyn G. The consultation game. Qual. Saf. Health Care. 2004;13; 415–416.

[9] Ciocci A., Viti A., Terzi A., Bertolacini L., The game theory in thoracic surgery: from the intuitions of Luca Paciolo to the operating rooms management. J.Thoracic Dis., 7,11 2015.

[10] Gonzalez Diaz J, Garcia Jurado I, Fiestras Janeiro M. G. An Introductory course on Mathematical Game Theory American Mathematical Society - Real Sociedad Matematica Espanola. 2010.

[11] Peters H. Game Theory - A Multileveled Approach. Springer; 2008.

[12] Pieri G, Pusillo L. Multicriteria partial cooperative games, Applied Mathematics, special issue on Game Theory. 2015;6:2125–2131.

[13] Maynard Smith J. Evolution and the Theory of Games. Cambridge: Cambridge University Press; 1982.

[14] Pusillo L. Evolutionary Stable Strategies and Well Posedness Property. Appl. Math. Sciences 2013 7.

[15] Cleveland C., Liao D., Austin R., Physics of cancer propagation: A game theory perspective. AIP Adv. 2,1,2012.

[16] McEvoy J W. Evolutionary game theory: lessons and limitations, a cancer perspective. Br J Cancer 2009;101:2060–1; author reply 2062–3.

[17] Djulbegovic B, Hozo I, Ioannidis J P. Modern health care as a game theory problem. Eur J Clin Invest 2015;45:1–12.

[18] Deisboeck T S, Zhang L, Yoon J, et al. In silico cancer modeling: is it ready for prime time? Nat Clin Pract Oncol 2009;6:34–42.

[19] Wu A, Liao D, Austin R. Evolutionary game theory in cancer: first step in prediction of metastatic cancer progression? Future Oncol 2015;11:881–3.

[20] Axelrod R, Axelrod D E, Pienta K J. Evolution of coocperation among tumor

cells. Proc Natl Acad Sci USA 2006;103: 13474–9.

[21] Moretti S, Patrone F, Bonassi S. The class of microarray games and the relevance index, TOP 2007;15:256–280.

[22] Fragnelli V, Pusillo L. Multiobjective Games for Detecting Abnormally Expressed Genes, Mathematics, special issue Mathematical Game Theory 2020; 8:3.

[23] Pusillo L, Tijs S. E-equilibria for multicriteria games, Annals of the International Society of Dynamic Games, by P. Cardialaguet and R. Cressman, Springer ed. 2012;12:217–228.

[24] Levaggi L, Pusillo L. Classes of multiobjective games possessing Pareto equilibria. Operation Research Perspectives 2017;4:142–148.

[25] Pusillo L. Vector Games with potential functions, Games 2017;8:1–11.

[26] Albino D, Scaruffi P, Moretti S, Coco S, Di Cristofano C, Cavazzana A. et al. Identification of low intratumoral gene expression heterogeneity in neuroblastic tumors by wide genome expression analysis and game theory. Cancer 2008;113:1412–1422.

[27] Capps D. John Nash, game theory, and the schizophrenic brain. J Relig Health 2011;50:145–162.

[28] DeLorenzo C., Papademetris X., Staib L., Vives K., Spencer D., Duncan J. Image-Guided Intraoperative Cortical Recovery Using Game Theory: Application to Neocortical Epilepsy Surgery. IEEE Trans.Med.Imaging, 29,2, 2010: 322–338.

[29] Bozma H I, Duncan J S. A game-theoretic approach to integration of modules. IEEE Trans. Pattern Anal. and Machine Intell. 1994;16:1074–86.

[30] Sanfey A G. Social decision making: insight from game theory and neuroscience. Science 2007;318:598–602.

[31] Chakraborty A, Duncan J S. Game-theoretic integration for image segmentation. IEEE Trans. Pattern Anal. and Machine Intell. 1999;21:12–30.

[32] Roth A E, Sönmez T S, Ünver U M. Kidney exchange. The Quarterly Journal of Economics, 2004;119(2):457–488.

[33] Roth A E, Sönmez T S, Ünver U M. Pairwise kidney exchange. Journal of Economic Theory, 2005;125(2):151–188.

[34] Villa S, Patrone F. Incentive compatibility in kidney exchange problems Health care management science. Springer; 2009.

[35] Myerson R. Game Theory. Harvard University Press, 3rd printing edition, 1997.